Party Wall Disputes: The Party Wall etc. Act 1996 and Beyond

Party Wall Disputes: The Party Wall etc. Act 1996 and Beyond

Legal Coherence and Dispute Management

Dr Laura Lintott, LLM PLP (BPP), MA (Oxon), PhD (Cantab), FCIArb

WILEY Blackwell

This edition first published 2025

Registered Offices
John Wiley & Sons, Inc., 111 River Street, Hoboken, NJ 07030, USA
John Wiley & Sons Ltd, The Atrium, Southern Gate, Chichester, West Sussex, PO19 8SQ, UK

For details of our global editorial offices, customer services, and more information about Wiley products visit us at www.wiley.com.

The manufacturer's authorized representative according to the EU General Product Safety Regulation is Wiley-VCH GmbH, Boschstr. 12, 69469 Weinheim, Germany, e-mail: Product_Safety@wiley.com.

Wiley also publishes its books in a variety of electronic formats and by print-on-demand. Some content that appears in standard print versions of this book may not be available in other formats.

Library of Congress Cataloging-in-Publication Data Applied for

Hardback ISBN: 9781394274963

Cover design by Dr Laura Lintott FCIArb and image by Alexander Lintott

Set in 9.5/12.5pt STIXTwoText by Straive, Pondicherry, India

Printed and bound by CPI Group (UK) Ltd, Croydon, CR0 4YY

C9781394274963_280525

Dedication

With thanks to those set out below for their comments, time, insights and inspirational and tireless support.

- My Family;
- Professor Martin Dixon KC (Hon), MA (Oxon), PhD (Cantab), FAcSS, Honorary Bencher of Lincoln's Inn, for his kind guidance throughout my PhD time and thereafter;
- The Earl of Lytton for his open-mindedness and kind thirst for discussion;
- Professor Renato Nazzini for his support and kind counsel;
- Nicholas Isaac KC for his kind wisdom;
- Tom Owen KC, MA (Cantab), for his friendship, gentle humanity and wit;
- Dr David Sawtell, MA (Cantab), MPhil (Cantab), MSc (KCL), PhD (Cantab), FCIArb, for reaching out when he did;
- Dr Paul Sayer for his warm patience and guidance;
- HHJ David Grant for his patience and approachability;
- Dr Alexander Wray for his sharp eye and friendship;
- As well a number of party wall surveyors, lawyers and friends.

Disclaimer: Any information provided in this book is informal only and is not binding on any person. You may therefore wish to seek your own legal advice on the matter. Any references to the law are up to 29 February 2024.

Contents

About the Author

Dr Laura Lintott, LLM PLP (BPP), MA (Oxon), PhD (Cantab), FCIArb is a Solicitor Advocate and International Arbitrator in London. She has extensive experience leading on complex international and domestic construction disputes. Laura is a Fellow of the Chartered Institute of Arbitrators and Visiting Fellow at King's College London where she joined the Centre of Construction Law & Dispute Resolution's Taskforce, which responded to the Law Commission's two consultations regarding Arbitration Act 1996 reforms reflected in the Arbitration Bill noted in the King's Speech in 2024. The Arbitration Act 2025 then received Royal Assent on 24 February 2025. Laura contributed a chapter to the book *The History of the Technology and Construction Court on its 150th Anniversary*, edited by Sir Peter Coulson and Dr David Sawtell. She is a frequent panellist/(podcast) speaker and publishes regularly in numerous journals and periodicals.

Foreword

Party wall practice and procedure occupies a complex niche in the spectrum of professional practice. Born of necessity after the Great Fire of London in 1666, it had become a tool of regulatory control by 1939 as part of the London Building Act. Part regulation to ensure best practice, part an ordering of ownership rights in areas of common interest and use, it incorporates notions of construction technology, an understanding of boundaries and their importance and resolution of disputes under a statutory umbrella.

And it was the transition from the local London Acts to a nationally applicable statutory process in which it was my great privilege to have a hand in. Translation from local metropolitan to national application was not entirely straightforward, not least because the conventions, assumptions and circumstances moved from a well-understood and to some extent standardised Inner London environment, to a far wider geographical area in which all manner of different situations might arise. Notable lacunae were a better definition of 'surveyor' and the loss of a sanction for failing to serve a notice before embarking on notifiable works, issues that remain for a future reform.

Coming to the matter as I do with the scars of parliamentary debate and procedure that accompanied my efforts to take the then Party Wall Etc Bill through its stages in the House of Lords where it began its passage, I must firstly pay tribute to those expert practitioners on whose shoulders I stood and most particularly those of the late John Anstey FRICS. Among parliamentary colleagues, the Earl of Kinnoull, Lord Dubs, Paul Beresford and Paul Everall deserve particular mention here.

But here and now in what I am certain will become a seminal treatise on the subject of party wall legislation and procedure it gives me great pleasure to preface the remarks of a legal expert in this area. Some 28 years after the Party Wall etc. Act 1996 came into force, it is indeed time for a review – a retrospective on what has happened since and forward thinking on where things might go from here.

Although I would never count myself as an expert on party walls, I can say with confidence that Dr Lintott has done the professions both of the law and of surveying in particular, as well as property owners, a great service in providing something of a stocktake, taking us to the heart of what ownership means in legal and practical terms, setting out the inevitable brokerage that professionals need to engage where the tectonic plates of title, use of land, statute and common law, easements, planning and construction codes, rights, liabilities and obligations all meet. In all of this the practicalities of give and take, essential where there are common (and potentially conflicting) interests with shared usage of physical features and defined proximity to neighbours, are not lost. In our increasingly densely developed built environments this continues to underpin one of the slickest and securest of property markets in the world. In terms of global competition and attraction this matters hugely to our national wealth and to the rules-based system. Dr Lintott's book is a very useful consolidation of matters of law, fact and dispute resolution options relevant to party wall disputes.

She is mindful of the reality that party wall practice is ultimately driven by mutual proprietorial and essentially practical considerations, by function of walls and other party structures, and within defined proximity, the effects on neighbours she points out that it relies hugely on the broader knowledge and competence of professionals and their assessment of practical implications of a building owner's proposed works and the logical means of resolving any differences that arise. Further she highlights that fomenting differences where they do not need to arise, failure to offer viable workarounds where these exist and standing on principle rather than on best and most cost-efficient practical outcomes are matters every professional should strive to avoid. She also appreciates the fundamentally facilitative nature of party wall procedures and practice and the importance of the informal tribunal created under the dispute resolution procedures of s.10.

Of course, there are many areas where party wall services could be improved: the definition of 'Surveyor' for instance; better and more specific training and possibly accreditation of practitioners and, as Dr Lintott suggests, express dispute avoidance provisions, and for lawyers, a more focussed and dedicated area of legal practice drawing together the many threads she identifies.

From my observation, it is fair to say that from the earliest days of its parliamentary process, there have been attacks on the fundamental premise, philosophy and process as represented by the Party Wall etc. Act 1996 and arguably of all its antecedent processes and precedents under the London Building Acts; some legal views held the Party Wall etc. Act 1996 to be offensive to the concept of title, that matters touching and concerning this core tenet of property ownership might be left to 'mere' surveyors – and these without any requirement for relevant knowledge and training at that! It was observed that appointed surveyors do not have any regulated functions in this respect though of course prior to 1997 scores of practitioners had decades of highly successful party wall practice behind them. On the other side were loft conversion and other specialist construction concerns which objected to the very idea that property owners might not have absolute and unfettered rights to do what they liked within the confines of their own property.

It may be that this consideration led to HS2 legislation dis-applying significant parts of the Party Wall etc. Act 1996 but seemingly the intention to reduce bureaucracy in public infrastructure schemes was also attended by a fundamental misunderstanding of what party wall procedure actually entailed.

Dr. Lintott ends by suggesting that a more focussed and integrated approach to the various interactions is now necessary; as one of those one time general practice surveyors who definitely learned much about party walls 'on the hoof' but having had a broad spectrum of property practice and experience, I can attest to the risks that a too narrow focus in professional expertise is indeed the potential enemy of something as interactive as party wall practice.

John Anstey was a wonderful raconteur of odd-ball situations he had encountered in professional practice, including the occasion when during foundation work, the neighbour's basement kitchen was inadvertently filled with fast setting concrete slurry! Fortunately, the surveying profession has risen to the challenge with some outstanding professionals in the field, many of whom recognise that accreditation is the way forward to secure public confidence, consistency of service and effective delivery.

In all this, practitioners would be wise to study carefully this timely reminder of just how many allied moving parts may be involved on the fringes of party wall practice which the Pyramus and Thisbe Learned Society and the Faculty of Party Wall Surveyors have been at pains to promote and reinforce. But ultimately it remains the case that the procedures under the Party Wall etc. Act 1996 are very narrowly defined so knowledge of scope and limitations and of the other areas the unwary practitioner might inadvertently stray into should at very least provide a timely route map to navigating the minefield.

January 2024 The Earl of Lytton

Preface

Throughout my career, I have worked as a construction disputes solicitor in private practice, which encompasses a rich mix of international arbitration, domestic litigation, mediation, expert determination and importantly advisory work to achieve dispute prevention, mitigation or settlement. In parallel, I wrote my PhD Thesis at the University of Cambridge (where I also took on teaching responsibilities), which formed the basis of this book. Additionally, I became a Visiting Lecturer and subsequently a Visiting Fellow at King's College London, where I am active at the time of writing this book. I teach on the MSc Construction Law and Dispute Resolution course, while working as a Solicitor Advocate and International Arbitrator in London. I also joined the Centre of Construction Law & Dispute Resolution's Taskforce, which responded to the Law Commission's two consultations regarding Arbitration Act 1996 reforms reflected in the Arbitration Bill noted in the King's Speech in 2024. The Arbitration Act 2025 then received Royal Assent on 24 February 2025. In 2024, I became a Fellow of the Chartered Institute of Arbitrators.

Over the years, both in my professional and private life, I have witnessed clients and friends experiencing adversity linked to party walls, with potential or actual related disputes arising out of these. Helping to mitigate such disputes has been a very rewarding experience and the aim of this book is to share with others the results of my PhD research and experience in this area.

Party wall disputes can escalate quickly. Once such a dispute starts, it is rare for the parties thereafter to be able to live harmoniously next to each other, especially after a court decision is served. This book therefore sets out a range of legal and factual issues, including and going beyond the Party Wall etc. Act 1996 (**PWA 1996**), to consider before and during a party wall dispute, including dispute resolution avenues that are available to the parties. These include the notice mechanism under the PWA 1996. They also include alternative dispute resolution methods in situations where: (i) the building owner fails to serve an appropriate party wall notice before starting the relevant works, as the PWA 1996 will not apply and the parties will need to go to court if a dispute arises; (ii) the building owner does serve the relevant notice under the PWA 1996, but the party wall award is appealed, and the parties end up in court as a result; or (iii) the building owner and adjoining owner enter into a contract choosing an alternative dispute resolution method to use should a dispute arise – if a dispute is settled, there is no dispute to apply the notice mechanism to under the PWA 1996 (you will see the relevant case law on this specific point in Chapter IV explaining why/how such an agreement can come about and what its effects are). In such situations, to save money, time and emotion, parties can avoid going to court or stop court proceedings by agreeing to pursue an alternative dispute resolution avenue such as settlement negotiation discussions between the parties (with or without lawyers), early neutral evaluation, mediation, medi-arb, arbitration, adjudication or expert determination.

As a disputes lawyer, I have seen how negative emotion can lead to the destruction of oneself and of others. On a more positive note, I have also witnessed how the conscious decision to collaborate, empathise with the other side and look beyond the immediate goals and emotions to reach a compromise can salvage seemingly unsolvable situations. Where there is a will, there is a way. It sometimes takes one person to turn things around, be it the legal representative, party wall surveyor or one of the parties or experts. That one person with a calm and positive attitude and openness to new solutions often has the power to motivate others to pause and think how to bring a gloomy dispute to an end, where everyone can feel like they have gained something. Let us all be that person.

Working in private practice alongside my academic and teaching endeavours has taught me to question long-established rules and principles, discuss and collaboratively challenge current practices with the intention of encouraging progress. Most things should be questioned from time to time, to enable practices to develop and improve outcomes.

Collaborative dialogue can speed up progress while instilling joy into people's lives. Prejudice, conscious and unconscious bias, fear of the unknown, personal interest and also fatigue all stand in the way of experiencing progress in a collaborative, exciting and accelerated manner. Everyone has a choice to listen and respectfully respond, support or try something else of their own. Success can grow exponentially through meaningful collaboration. Improvement is a never-ending process and should be encouraged. Attempts at improvement may not always be perfect but without such attempts, there is no progress. The main thing is to keep on trying.

Working on this book has motivated me to collaborate with lawyers, party wall surveyors, academics and friends, to ask questions, reflect on some current set views and procedures and seek new helpful approaches. I have been fortunate to meet enlightened lawyers and party wall surveyors with brilliant insights and open minds. I do not have all the answers; therefore, I would like to encourage readers of this book to continue going with time and keep developing solutions that can help those in difficult situations prevent, mitigate, settle or manage any party wall disputes they find themselves in.

The evolution of the law cannot be stopped. It is fascinating to watch its transformations and rewarding to be part of the process when it takes positive turns. This book recognises the importance and value that both lawyers and party wall surveyors bring to the table and embraces cooperation between the two professions which is in the best interest of the clients who need their help.

This book also emphasises the need to think contextually, to take into consideration areas of law and fact that go well beyond the PWA 1996, in order to be well informed and to choose a path forward that may not have been apparent at first glance. Knowing when to stop escalating a dispute that is disadvantageous from commercial, emotional and time perspectives for either/any of the parties involved is a true skill and is not easy when embroiled in a dispute. This is where third parties can help (lawyers, party wall surveyors, mediators etc.).

This book does not replace the need to consult a lawyer. If you are reading this book as someone who is facing a party wall dispute, I encourage you to seek legal advice in good time, which may prevent having to spend much more substantial sums later down the line.

I do hope you will enjoy reading this book and that it will give you some comfort should you be facing a party wall dispute yourself. I also hope this book will spark more curiosity in this highly relevant area of law. Most importantly, I wish you all the success in preventing or amicably resolving party wall disputes, be it in your private or professional life.

The electronic version of this book will feature colour diagrams.

All the best!

Dr Laura Lintott, LLM PLP (BPP),
MA (Oxon), PhD (Cantab), FCIArb

Glossary of Definitions

1939 Act London Building Act 1930 and London Building Acts (Amendment) Act 1939, together

39HP 39 Headfort Place, London SW1X 7DE (*Wellington Properties Limited v The Trustees of the Will of the Second Duke of Westminster, Grosvenor Estate Belgravia*) [2018] EWHC 3048 (Ch)

ADR Alternative dispute resolution

ANLA 1992 Access to Neighbouring Land Act 1992

Approved Document E Guidance issued by the Secretary of State for the purposes of giving practical advice as to the requirements of Schedule 1 to, and Regulation 7 of, the Building Regulations 2010 (SI 2010/2214) for England and Wales (HM Government, 'Resistance to the passage of sound – Approved Document E' (2003 edn incorporating 2004, 2010, 2013 and 2015 amendments))

Bill Property Boundaries (Resolution of Disputes) Bill [HL] 2019-20

BRE Building Research Establishment

BRE Guide Building Research Establishment guidelines

Building Regulations Building Regulations 2010 (SI 2010/2214)

Convention European Convention for the Protection of Human Rights and Fundamental Freedoms (Council of Europe, Rome, 4 November 1950)

COPA 1974 Control of Pollution Act 1974

CPR Civil Procedure Rules (The White Book Service 2024 (Sweet & Maxwell, 2024))

DCLG The Department for Communities and Local Governments

DCLG's Explanatory Booklet DCLG's explanatory booklet (with the latest version from May 2016) on the PWA 1996, https://www.gov.uk/government/publications/preventing-and-resolving-disputes-in-relation-to-party-walls/the-party-wall-etc-act-1996-explanatory-booklet

EPA 1990 Environmental Protection Act 1990

First Protocol First Protocol to the Convention

FPMA 1774 Fires Prevention (Metropolis) Act 1774

HHJ His or Her Honour Judge

HRA 1998 Human Rights Act 1998

HS2 Legislation High Speed Rail (London – West Midlands) Act 2017 and High Speed Rail (West Midlands – Crewe) Act 2021

LRR 2003 Land Registration Rules 2003/1417

LRA 2002 Land Registration Act 2002

LPA 1925 Law of Property Act 1925

Order Party Wall etc. Act 1996 (Electronic Communication) Order 2016 (SI 2016/335)

PA 1832 Prescription Act 1832

Protocol A protocol issued by senior members of the legal and surveying professions, in connection with disputes between neighbours and the location of boundaries between their respective land (S Tozer, G Fetherstonhaugh QC, J Karas QC, N Cheffings and M Ditchburn, 'Protocol for disputes between neighbours about the location of their boundary (The Boundary Disputes Protocol)' (2017), https://www.propertyprotocols.co.uk/the-boundary-disputes-protocol)

PWA 1996 Party Walls etc. Act 1996

RICS Royal Institute of Chartered Surveyors

VSC Vertical Sky Component

WP Without prejudice

WPSATC Without prejudice save as to costs

1

Introduction

1.1 The Aim of This Book

Party walls are like salt. Many of us may not be consciously thinking about them, but most of us are affected by them. This book is meant to be a connecting point for members of the public, party wall surveyors, lawyers, academics and anyone interested or involved in party wall disputes. The aim is a collaborative approach with one goal in mind – to help parties in party wall disputes save emotion, time and money.

Party walls affect most of those who live in residential buildings, work in commercial buildings and/or manage commercial properties. Party wall disputes can be unnecessarily costly (e.g. performing unapproved works, having to tear a wall down, expenses linked to litigation, etc.) and time-consuming due to the lack of a clear structure on how to deal with them and a lack of awareness of which areas of law and fact are relevant to them as the Party Wall etc. Act 1996 (**PWA 1996**) is only one piece of the puzzle. This can result in a negative impact on individuals, families and companies, causing them emotional and financial hardship, particularly where parties have been neighbouring landowners for decades and where breakdown of their relationship can leave one or both parties with no other choice but to sell their respective properties (possibly at a lower than market rate to expedite the sale). Where party wall disputes end up in court, this is often unnecessary and places a burden on the court system.

In an ideal world, parties to party wall disputes would adhere to the statutory provisions of the PWA 1996 whereby the building owner serves the appropriate party wall notice before starting any works affecting the adjoining owner in scope of the PWA 1996. This ensures that both the building owner and the adjoining owner can enjoy the rights conferred by the PWA 1996. The advantage is then that the parties can benefit from one of the main purposes of the PWA 1996, namely resolving their party wall dispute under the PWA 1996 outside of court. Should a dispute arise, parties would appoint a party wall surveyor(s) to render an award, hopefully resolving such a dispute. While such an award can be appealed in court, parties still have the option to agree to try and negotiate a settlement by way of alternative dispute resolution (**ADR**), for example through mediation. In such a case, parties would agree to suspend court proceedings, and if they agree on a settlement, court proceedings can be formally brought to an end. However, the option of ADR is open at any point in time,

Party Wall Disputes: The Party Wall etc. Act 1996 and Beyond – *Legal Coherence and Dispute Management*, First Edition. Dr Laura Lintott.

prior to as well as during court proceedings. The parties even have the possibility by agreement to ensure that the dispute resolution procedure in Section 10 of the PWA 1996 is not engaged and pursue a form of ADR instead of the mechanism prescribed under the PWA 1996.[1] However, this is not always appropriate, and parties should consider seeking impartial advice from lawyers as well as party wall surveyors prior to taking such a step.

Where the building owner fails to serve an appropriate party wall notice, the PWA 1996 will not apply. This is where again ADR, as an alternative to litigation, is a desirable way forward, as it can preserve or repair relationships between landowners who are in party wall disputes, who are by definition literal next-door neighbours, and save the parties' time and money. Party wall disputes tend to be particularly painful and can often be disproportionately expensive because feelings can run so high. Agreement is therefore to be encouraged, for example, with the assistance of a party wall surveyor, mediator or adjudicator. There is a repetitive pattern of how party wall disputes can arise, how they may be mismanaged by not complying with the relevant legislation and failing to seek guidance from surveyors and lawyers – resulting in a negative impact on people's wellbeing and lives. The repetitive pattern of party wall disputes, however, gives rise to a solution – systematisation of recurrent issues and recognition of warning signs prior to disputes arising. This in turn gives rise to the potential for dispute prevention if well managed and the provision of a selection of potential dispute resolution solutions.

The purpose of this book is to take a holistic view of party wall matters. The aim is therefore to not only analyse the PWA 1996 but also look far beyond it – to bring together those areas of law and fact that are most relevant to, and frequently encountered in, party wall disputes; provide a more comprehensive understanding of party wall disputes; and help resolve such disputes in a practical and effective way.

This book analyses points of friction between owners or occupiers of neighbouring properties linked by party walls. At first glance, issues connected to party walls would seem to be regulated by the PWA 1996. While the PWA 1996 focuses on party walls and attempts to provide a notice dispute resolution mechanism in relation to disputes arising out of them, it is very narrow in its scope and does not consider crucial areas of law and fact. It also is not helpful where the building owner fails to trigger the notice dispute management mechanism created by the PWA 1996 (statutory breach). Additionally, the PWA 1996 does not compel parties to consider the range of other available out-of-court dispute resolution mechanisms, which are further discussed in this book, which provide cost- and time-efficient alternative avenues for disputants. ADR is supported by courts.

Having said that, the PWA 1996 does contribute with some positives, including the unification of previous legislation now applied nationwide and, in particular, the notice dispute resolution mechanism for the parties to attempt to resolve disputes out of court. This addresses disputes where an adjoining owner explicitly does not agree with the building owner's proposed works or does not respond to a notice by the building owner relating to such works, or where the building owner does not respond to the adjoining owner's counter-notice. The surveyor's award then focuses on works that fall under the PWA 1996. The award does not, however, go beyond the PWA 1996. Effective use of the PWA 1996 can

---

1 *Dillard v F & C Commercial Property Holdings Ltd* [2014] EWHC 1219 QB.

be achieved only if done in the context of other legal and factual areas and with the right collaborative guidance from party wall surveyors (within their PWA 1996 jurisdiction) and legal professionals.

One of the main issues is that there is a need for a contextual approach, which is rarely undertaken, that recognises the multifaceted nature of party wall disputes. The dispute resolution process dealing with the layered and sometimes intertwined areas of law and fact behind party wall disputes can be highly contentious, if not acrimonious. Courts are unnecessarily burdened with party wall disputes that could and should be resolved out of court.

Another issue is that a layperson may conclude that, in applying the PWA 1996 themselves, the PWA 1996 will provide a sensible and safe guidance for the party to deal with a party wall dispute on their own, without consulting a legal professional. Unfortunately, this is a misleading impression that the PWA 1996 can create, which can lead to a situation detrimental to the parties trying to solve party wall disputes on their own rather than referring to legal specialists in this area (however, there are not many party wall specialists available on the legal services market at the time of writing this book) and by failing to consider points of law and fact crucial to their dispute. In addition, the profession of party wall surveyors is not regulated, which means that there can be a great spectrum in terms of quality of any party wall award under the PWA 1996. Keep in mind that party wall disputes can be affected by issues outside the scope of the PWA 1996 (and therefore outside of a party wall surveyor's jurisdiction). Therefore, advice from both a party wall surveyor and lawyer is something parties should consider. The party wall surveyor can render an award under the PWA 1996 if appointed to do so. However, a legal specialist can advise on the PWA 1996 as well as further issues, which are discussed in this book (such as other pieces of legislation, tortious matters or proprietary rights – and how these areas coexist with the PWA 1996), how to appeal an award in court, and which form of ADR is appropriate in any given case and strategically when. Party wall surveyors should explain the limitations of their jurisdiction to the parties. Ideally, party wall surveyors and lawyers will assist the parties, in spirit of collaboration between the two professions, to reach an amicable solution between the disputants.

A fascinating aspect of the PWA 1996 is that it was introduced as a Private Member's Bill in the House of Lords by the Earl of Lytton. Considering that Private Member's Bills seldom materialise into actual legislation, this is a substantial achievement. This book therefore also explores how and why sufficient parliamentary time was found for the measure to pass into law.

As noted above, party wall disputes often touch on other areas of statute law that can impact each other (for example, the Access to Neighbouring Land Act 1992, the Land Registration Act 2002 or the Human Rights Act 1998), property rights (such as rights of way, rights of support, drainage rights or rights linked to easements) and tort (such as nuisance related to noise and vibration, or breaches of rights to light). In addition to areas of law, there are factual matters that also need to be considered in the context of party wall disputes and include structural issues (such as cracks in party walls, subsidence, underpinning, thickening, raising, repairing, cutting into a party wall, increasing or decreasing its height, exposing a party wall or demolishing it and fully rebuilding it).

There is a need for a new approach to party wall disputes, as they should no longer be viewed as peripheral issues linked to property, construction or commercial disputes and projects. There is a lack of a specialism primarily focusing on party wall disputes as a valid and complex legal area of its own merit. Party wall disputes require a multifaceted approach.

As the range of issues linked to party walls and related dispute management is currently fragmented, there is a necessity to connect the different areas of law and fact related to party walls, analyse how and why they are relevant and determine how disputes emanating out of them can be better managed, ensuring that relevant legal principles are assessed and adhered to, giving clarity to courts and relevant parties. Providing such clarity and mapping out the area of party wall matters, bringing in different elements of the law and fact, means creating a separate party wall discipline of its own, which is one of the aims of this book.

This book also presents dispute management and resolution options, including a strategy analysis to be applied prior to a dispute arising, as well as settlement negotiation, surveyor's award, mediation, expert determination, litigation, etc. This book advocates for a pragmatic cost- and time-effective dispute resolution approach, out of court where possible (considering the common disproportionality between the value in dispute and costs/time connected to litigation in party wall disputes).

The new approach in this book calls for a reform of the party wall dispute resolution system. By setting out links between different legal areas as well as factual issues, it becomes possible to systematise issues surrounding party walls so that they become predictable and therefore more manageable. This in turn results in greater efficiency and effectiveness in party wall dispute prevention, management and resolution. This book provides different perspectives when it comes to party wall disputes and, in some cases, even offers a choice of which areas of law/issues to prioritise to achieve a more favourable dispute outcome.

Parties do not always have to resort to court proceedings if they are not happy with an award under the PWA 1996 or circumstances linked to other areas of law and can try other options to address the dispute. Chapter IV on dispute resolution analyses dispute management under the PWA 1996 but also practical aspects of party wall disputes and other dispute resolution avenues (ADR). It gives solutions and options as to how parties can resolve party wall disputes outside of court. Factors reflected include awareness of and sensitivity towards relationships between neighbours, and the time and cost involved in party wall disputes, which are all valid considerations for parties to party wall disputes when deciding how to manage and address such disputes.

Although the complexity of party wall disputes is addressed in this book, the simple and most common reason as to why party wall disputes arise is often the basic right, need and emotion behind preserving the landowner's or occupier's territory and to not negatively impact on the adjoining land.

As noted above, the purpose of this book is to carve out a distinct party wall discipline of its own, a subcategory of neighbourly matters. This is to offer coherence, providing an overview of a range of factors that need to be taken into consideration when it comes to party wall disputes. It is to serve as an inspiration to investigate alternative methods of dispute resolution to preserve neighbourly relationships and save cost and time.

Some of the case law discussed in this book predates the PWA 1996. This is because the PWA 1996 reflects its legislative predecessors, with only a few significant changes, as the main aim was to extend the legislative application to the whole of England and Wales.

The case law preceding the PWA 1996 assists in giving a fuller context/picture of the area of party walls when touching on other areas of law and/or interpretation of the PWA 1996. Case law predating when the PWA 1996 was enacted also provides a rich spectrum of examples showing how the law around party walls has developed.

> *The forerunner of the Party Wall Act 1996 was the London Building (Amendment) Act 1939. Between s.46 and s.56, the 1939 Act contained detailed provisions whereby a building owner who wanted to carry out works in or around a party structure was obliged to give notice in writing, and thereafter any differences between the owners would be resolved in the first instance either by one surveyor, or by a surveyor appointed by each party and a third surveyor. The 1939 Act was considered to have worked well in London, and eventually it was decided to extend the regime across the country. Hence the Party Wall Act, which mirrors the essential features of the 1939 Act (although some of its provisions are different).*[2]

1.2 Chapter Breakdown

The book is divided into chapters ranging from an introduction to the topic, through identifying the issues linked to party walls and grouping them by relevant type of law and fact, to dispute resolution and management options. The analysis sets out the topic of party walls and focuses on the immediate issues emanating from party wall disputes. It investigates dispute management and resolution options. It shows the wide spectrum of legal and factual issues surrounding party walls and the appropriate dispute resolution management, avenues, process and solutions.

Chapter I is an introduction to the book, formulating a party wall area of its own, providing a systematisation of most pertinent issues and assisting in efficient and effective party wall dispute resolution. It explains how this book is connecting different areas of law and fact related to party wall disputes. It shows how it aims to help members of the public to save money and time in such disputes by strategic preparation and planning leading to early settlement. The introduction sets out the benefits of this book to a) practitioners by explaining the need to look beyond and above one specific area of law/fact and the importance of contextual thinking; b) academics by uncovering an independent new multifaceted area of party walls and related disputes; and c) the public, by suggesting where to get the right information and advice from to pragmatically prevent or solve party wall disputes (and not to be misled by considering the PWA 1996 as the only legislation to refer to).

Chapter II sets out what party walls are in the first place in connection with different pieces of legislation and case law. It also discusses the way in which the PWA 1996 distinguishes between party walls, party fence walls and party structures, as well as the main types of party walls. This is illustrated with examples from the Department for Communities and Local Government, '*Party Wall etc. Act 1996: Explanatory Booklet*'[3] that was produced

2 *Power v Shah* [2023] EWCA Civ 239, 9.
3 (May 2016).

to assist with interpreting the PWA 1996. The chapter also touches on excavations and special foundations.

Chapter III deals with the PWA 1996. It focuses only on issues linked to party walls to the extent to which the limited framework provided by the PWA 1996 applies. It gives a historical overview of the legislative developments leading to the inception of the PWA 1996 and explains the main areas covered by the PWA 1996 as well as rights and obligations of the parties connected to party walls.

Chapter IV analyses the different dispute resolution avenues that are most relevant to party walls. It explains the notice dispute resolution mechanism in relation to party wall disputes under the PWA 1996. Then the chapter proceeds to analyse additional forms of dispute resolution as to party walls that are available to parties. The chapter also discusses the strategy behind the decision-making process as to which dispute resolution avenue is necessary or possible and approaches that parties to party wall disputes may find most beneficial in the context of their circumstances in terms of emotion, cost and time. The aim is to show what options are available to those affected by current and/or potential disputes arising out of issues linked to party walls.

Chapter V looks at other areas of statute law that had or have an active interface with the PWA 1996. This includes the Access to Neighbouring Land Act 1992 (**ANLA 1992**). The ANLA 1992 is relevant to this book as it gives limited right of access to a neighbour's land to carry out 'basic preservation works'. The question is whether this is relevant to party walls. Prior to the ANLA 1992 having been passed, adjoining property owners had virtually no right to go onto their neighbour's land unless an express easement had been granted, such as a right to maintain drains, pipes and wires. The ANLA 1992 is therefore discussed in this book as well, and whether it dovetails with the PWA 1996.

In connection with this, practical considerations are analysed that are relevant to party walls, such as scaffolding and crane oversail and their potential interactions with party walls.

The Land Registration Act 2002 in the context of boundary determination is also discussed. The chapter explains what this involves, what the general boundaries rule is and the rules for fixing boundaries under the two Land Registration Acts. In addition, in the context of boundary determination, Chapter V explains what role the Law of Property Act 1925 plays. Tangible case law examples are used to show the connection between the aforementioned issues and party walls.

The chapter analyses the Property Boundaries (Resolution of Disputes) Bill (**Bill**). While there is no certainty that the Bill will progress through Parliament, it is still relevant to this book as the Bill aims to set out a clear dispute resolution structure in relation to the location or placement of boundaries and private rights of way regarding an estate in land title without having to go to court. It is a piece of legislation modelled on the PWA 1996.

Chapter V also discusses the Human Rights Act 1998 together with the European Convention of Human Rights as it touches on the landowner's or land occupier's right to a fair trial and the right to respect for private and family life, both relevant in the context of party walls.

Chapter V further considers the Control of Pollution Act 1974 and the Environmental Protection Act 1990, from which local authority environmental departments derive powers to deal with certain issues that can be linked to party walls, such as potential nuisance, dust and deposits from construction sites.

The Crossrail Act 2008, the High Speed Rail (London – West Midlands) Act 2017 and the High Speed Rail (West Midlands – Crewe) Act 2021 are also touched upon as they, in some respects, override the PWA 1996.

All of these pieces of legislation are relevant to party walls as they either interact with, amend, complement or coexist in parallel with the PWA 1996.

Chapter VI analyses proprietary rights in land (with a focus on easements). Proprietary rights govern parties' ability to use and enjoy both land in their physical possession and land in the physical possession of others. This is highly relevant to, and can be interlinked with, party walls.

Chapter VI focuses primarily on the rights set out below.

a) The right of way with which party walls can interfere/coexist.
b) The right of support (also covering weather protection and excavations), highly relevant to party walls, which can either require or provide support. This in turn leads to related rights of the landowners/occupiers with an interest in the relevant party walls.
c) Drainage rights.
d) Whether there are ancillary rights or obligations to repair a party wall and by whom.

Chapter VI highlights a number of cases together with clear images of situations where proprietary rights in land affect party walls and vice versa.

Chapter VII explores further aspects of relevance to party walls from the tortious perspective. It initially explains the meaning of common law nuisance before analysing examples of nuisance relevant to party walls, such as noise nuisance (including control of construction noise, soundproofing and vibration) and interference with rights to light. Examples of implications for party walls and related disputes stemming from tort are illustrated by way of case law.

Chapter VIII focuses on factual issues related to party walls and structural issues specifically. These include, for example, cracks in party walls, subsidence, underpinning, thickening, raising, repairing, cutting into party walls, increasing or decreasing party wall height, exposing party walls or demolishing these and fully rebuilding them.

Chapter IX is the conclusion to this book. It provides a synthesis explaining how the different chapters of the book fit together and what inference this leads to. This includes an assessment of the legal coherence of how party wall disputes are dealt with and the best options in terms of party wall dispute management.

2

What Is a Party Wall?

2.1 Introduction

Before one can analyse party wall disputes, and the related legal coherence forming the backbone behind the topic of party walls and related disputes, the first step is to identify what 'party walls' are.

Party walls can obstruct a view, limit a source of light, impede a neighbour's airspace or have an impact on the support of neighbouring land or property. Party walls need to be repaired from time to time, so the question arises as to who needs to repair them. This in turn leads to issues about whether it is possible to require a person to repair a party wall. Parties need to be clear on their rights to be able to access the relevant party walls. If a party attempts to access a party wall and wrongfully interferes with the other party's possessory rights in its land, this can lead to trespass. Overgrown roots of a tree can interfere with a party wall. Works on a party wall can cause construction noise and vibration. Alternatively, party walls in buildings may not be insulated enough to prevent certain noises from carrying from one room to another. A party wall is not necessarily visible in its entirety. It can be linked to the ground or to a structure. It can be linked to the ground below the land surface, and it can be impacted by ground conditions.

From the above examples, it is clear that party walls can be at the root of many disputes linking statutory, proprietary, tortious and structural matters. This is simply because of the highly contentious nature of the disputes linked to party walls. When the parties are evaluating which dispute resolution avenue they could or should take, this depends on their personal preference, willingness or necessity to preserve the relationship between the neighbouring landowners, cost and time available and need for clarity, certainty and a binding outcome.

Party Wall Disputes: The Party Wall etc. Act 1996 and Beyond – *Legal Coherence and Dispute Management*, First Edition. Dr Laura Lintott.

2.2 Definitions of 'Party Wall'

When it comes to defining what exactly the terminology 'party wall' means, there is no one clear-cut definition.[4] Before a legal estate in tenancy in common was abolished by the Law of Property Act 1925, in relation to a wall that was owned in common by different owners, each of the owners had generally the right to demolish the wall as long as the intention was to rebuild it again and to do the required repairs and underpinning.[5] However, case law made it clear that the provisions of the party wall legislation superseded such common law rights.[6] In common law, the terminology 'party wall' is not defined and does not have an exact meaning.

According to Megarry & Wade, since 1925,[7] the term 'party wall' may mean any one of the below.[8]

> *(i) a wall divided longitudinally into strips, one belonging to each of the neighbouring owners; or*
> *(ii) a wall divided as in (i), but each half being subject to an easement of support in favour of the owner of the other half;*
> *(iii) a wall belonging entirely to one of the adjoining owners, but subject to an easement or right in the other to have it maintained as a dividing wall.*
>
> *Prior to 1926 there was a fourth category. There could be a party wall of which the two adjoining owners were tenants in common. The disadvantage of such a wall was that either owner could insist upon partition. Had special provision not been made by statute, all party walls in this category would have become subject to a trust for sale after 1925. It was provided therefore that after 1925 all party walls of this kind, whether created before 1926 or after 1925, should be deemed to be severed vertically, and that the owner of each part should have such rights of support and use over the rest of the wall as were requisite for giving the parties rights similar to those which they would have enjoyed had they been tenants in common of the wall. The practical effect of this provision was to translate all party walls of this kind into the second category listed above.*

To elaborate on the terminology in the different categories, 'longitudinal division into two strips' in (i) above means that the ownership of the party wall is split longitudinally between the two landowners. However, cases of longitudinal division of this kind are rare. The issue with this category is that neither owner has any right of lateral support from one another. This can lead to either owner removing their half of the wall. While having to act with reasonable care, the result may be a structure that is not capable of standing on its own.[9]

The effect of the Law of Property Act 1925 (Section 38) was that a party wall is not owned in common but longitudinally where each of the parties has an easement of support over each other's property. While party walls fell under the law of co-ownership or easements of support, they had to be excluded from the statutory trust for sale that was imposed in cases

4 *Kempston v Butler* (1861) 12 I.C.L.R. 516, 526.
5 *Standard Bank of British South America v Stokes* (1878) 9 Ch. D. 68.
6 ibid.; *Selby v Whitbread & Co* [1917] 1 K.B. 736.
7 Law of Property Act 1925, 38.
8 Megarry & Wade, *The Law of Real Property*, 9th edn, (Sweet & Maxwell Ltd, 2019) 29–044.
9 ibid. 29–045.

of co-ownership after 1925.[10] There are easements that can affect party walls, which is why a separate Chapter VI is dedicated to easements in connection with party walls.

According to *Gale on the Law of Easements*:[11]

> *A wall could (and still can) be a party-wall up to a certain point, namely, so far as it divides two buildings of unequal height, and an external wall above that point; and a pilaster or portico, or a fascia, which appears to form an integral portion of one house, may be parcel of and pass on a conveyance of another house. The raising of a party-wall by one part owner without the consent of the other is a violation of that other's right of ownership and possession of his half, and is not less so because a private Act provides that it shall be lawful for the owner or part owner of any party-wall to raise it, provided that the wall when raised will be of the substance required by any byelaw.*

The above statement further shows that where a wall is a 'party wall', it is linked to certain rights and obligations that parties to the party wall must abide by.

As opposed to the definitions addressed above, which focus on ownership, when it comes to the definition of what a 'party wall' is under the PWA 1996, this definition does not relate to ownership but to the function of the wall. The PWA 1996 applies to England and Wales, and some of its provisions were adopted from legislation specific to Inner London and other parts of the country, which were in existence for a number of years prior to the PWA 1996.[12] The PWA 1996 regulates the construction and repair of party walls as well as related dispute resolution within the realms of the PWA 1996.[13]

Section 20 of the PWA 1996 states:

... *'party wall' means –*

> *(a) A wall which forms part of a building and stands on lands of different owners to a greater extent than the projection of any artificially formed support on which the wall rests; and*
> *(b) so much of a wall not being a wall referred to in paragraph (a) above as separates buildings belonging to different owners.*

As *Keating on Construction Contracts* says, a party wall always forms part of one or more buildings and where a wall does not stand on the land of both owners, such a wall will constitute a party wall only to the extent that it separates two buildings (as per the definition of Section 20(b) of the PWA 1996).[14]

By way of example, where a wall separates two terraced houses up to the height of two storeys (out of eight), however the wall is that of one building at the next (third) storey, for level three, it will not be a party wall. It would be a party wall only if it stood on the lands of both owners. Equally, where a wall separates houses for the full depth of one house but then carries on as the external wall of the other house, it will not be a party wall from the point of the end of the first house. Again, it would be a party wall only if it stood on the lands of both owners.[15]

---

10 ibid. 29–046.

11 *Gale on the Law of Easements*, 21st edn, (Sweet & Maxwell Ltd, 2020) 11–01.

12 ibid. 29–049.

13 Megarry & Wade (n 8) 29–043.

14 11th edn, Mainwork & 3rd Supplement, (Sweet & Maxwell Ltd, 2023) 16–086.

15 *London, Gloucestershire & North Hants Dairy Co v Morley & Lanceley* [1911] 2 K.B. 257; *Drury v Army & Navy Auxiliary Co-operative Supply* [1896] 2 Q.B. 271.

Notably, the PWA 1996 does not abolish fabric rights, meaning rights of ownership and support. On the contrary, it places an obligation on the building owner to provide protection from weather in a case where the building owner exposes a party wall to the elements.[16] There can be temporary interference with, for example, support that the party wall provides during construction works.[17]

Section 20 of the PWA 1996 distinguishes between:

a) Party walls – Walls forming part of a building and standing on lands of different owners to a greater extent than the projection of any artificially formed support on which the walls rest and walls separating buildings belonging to different owners.
b) Party fence walls (or boundary walls) – Walls, which are not part of buildings, stand on land of different owners and are used or constructed to be used for separating such adjoining. However, party fence walls do not include walls that have been built on the land of one owner where the artificially formed support projects into the land of another owner.
c) Party structures – Party walls and also floor partitions or other structures separating buildings or parts of buildings approached solely by separate staircases or separate entrances.

All three types of walls/structures are further elaborated on below.

Party walls envisage some form of solid structure, their purpose being to divide properties of different landowners or occupiers. The Department for Communities and Local Governments (**DCLG**) has issued an explanatory booklet (with the latest version from May 2016) on the PWA 1996 (**DCLG's Explanatory Booklet**). It provides more detailed explanatory information and guidance provided by the government on party wall issues with illustrations showing the differing main scenarios.

The PWA 1996 is silent on what materials party walls might consist of. According to DCLG's Explanatory Booklet, a party wall needs to be of a more solid nature whereby materials such as brick or cement would qualify as being sufficient to form a party wall. However, this is not stated anywhere in the PWA 1996. Where the PWA 1996 is not clear or is mute on a point relevant to party walls, it is a good idea to check DCLG's Explanatory Booklet. While DCLG's Explanatory Booklet may not answer every question, nor does it always give perfect answers, it does provide some guidance, elaborates on the PWA 1996 with practical examples and assists in interpreting the PWA 1996.

In conjunction with the type of materials party walls should be made of, building regulations have existed in one shape or another since cities emerged. This stems from the fact that, when buildings collapsed or were on fire, the consequences could be potentially catastrophic. Governments therefore have traditionally tried to intervene to prevent loss of life and property.[18] Looking at fire safety legislation, this has reacted from a historical perspective to fatal or catastrophically damaging fires, including the Great Fire of London in 1666, after which the London Rebuilding Act 1667 decreed: '[t]hat all the outsides of buildings be henceforth made of brick or stone'. This was to replace timber. It was envisaged for party walls and external walls to be one to two bricks in thickness. In addition,

16 PWA 1996, s 2(2)(n).
17 S Bickford-Smith and C Lamont, 'Party Wall etc. Act 1996 – 10 years on' (Landmark Chambers, 2007), 6.
18 H Carr, 'Grenfell Tower and the failure of building and fire safety regulations introduction' (2017) 20(5) *Journal of Housing Law* 110–115.

streets were to be of sufficient width to prevent fire spreading across them.[19] District surveyors were responsible for enforcing the regulations, and anyone who erected a building in contravention of these regulations had to face a criminal offence charge. The relevant building would then be destroyed, as it would be deemed to be a common nuisance.[20]

Before analysing the legal relationships linked to party walls, the question is who are the parties that claim ownership over party walls. Section 20 of the PWA 1996 defines the terms set out below.

a) An 'adjoining owner' and 'adjoining occupier' respectively mean "*any owner and any occupier of land, buildings, storeys or rooms adjoining those of the building owner and for the purposes only of section 6 within the distances specified in that section;*"
b) A 'building owner' means "*an owner of land who is desirous of exercising rights under this Act;*"
c) An 'owner' is a term that is non-exhaustive and includes,[21] as per section 20 of the PWA 1996:
   (a) a person in receipt of, or entitled to receive, the whole or part of the rents or profits of land;
   (b) a person in possession of land, otherwise than as a mortgagee or as a tenant from year-to-year or for a lesser term or as a tenant at will; and
   (c) a purchaser of an interest in land under a contract for purchase or under an agreement for a lease, otherwise than under an agreement for a tenancy from year-to-year or for a lesser term.

The building owner needs to establish whether the object in question is a party wall, a party fence wall or a party structure under the PWA 1996 in order to find out whether the PWA 1996 applies and, if so, what requirements the building owner has to meet in order to comply with the PWA 1996.

As for adjoining occupiers (including a large class of periodic tenants) (**Adjoining Occupiers**) that do not qualify as owners, they have rights of compensation under Section 7(2) of the PWA 1996 and the right to receive notice prior to property being entered under Section 8 of the PWA 1996. However, Adjoining Occupiers do not have the right to be heard in deliberations that lead to an award determining what works are to be carried out.[22]

The two main distinguishing factors are a) the position of the party wall relative to the land's boundary; and b) if any structures are attached to it or are part of it. In addition, the building owner has to identify who owns the wall, which may not be always clear due to boundary issues. These are further discussed in Chapter V on boundary determination.

The PWA 1996 addresses three types of works specified in Sections 1, 2 and 6 of the PWA 1996.[23]

1) new building on line of junction (Section 1);
2) works to existing party structures (Section 2); and
3) excavations within three or six metres of neighbouring buildings or structures in specific circumstances under the PWA 1996 (Section 6).

---

19 M Edwards, 'Fire safety in tall buildings – Part 1: tall, super-tall and mega-tall' (2017) 33(7) *Construction Law Journal* 442–445.
20 Carr (n 18) 110–115.
21 *Keating on Construction Contracts* (n 14) 16–090.
22 S Bickford-Smith, D Nicholls, and A Smith, *Party Walls Law and Practice* 4th edn (LexisNexis, 2017), 2.11.
23 *Mohamed & Lahrie v Antino & Stevens* (CCCL, 13 December 2017); N Isaac, *The Law and Practice of Party Walls* 2nd edn (Property Publishing, 2020) 3-01-5-65.

The main types of party walls include:[24]

a) Type A Wall – A wall standing on the land of two owners to a greater extent than just projecting foundations; and
b) Type B Wall – The part of a wall standing on the land of one owner that separates the buildings of two property owners.

Examples of Type A Walls are shown in Diagrams 2.1, 2.2, 2.3, 2.5 and 2.6 and Type B Walls in Diagrams 2.1 and 2.4. The distinction between a Type A Wall and Type B Wall is vital for the following reasons:

i) In the case of a Type A Wall, the wall can be part of one building or two (or more) separate buildings. These types of walls usually form part of terraced or semi-detached houses. The location of the wall's foundations should not be taken into account when investigating whether or not the wall stands on a boundary.
ii) In the case of a Type B Wall, it is a wall which stands entirely on the land of the building owner and the adjoining owner has an enclosed building. The Type B Wall has to be enclosed on both sides.

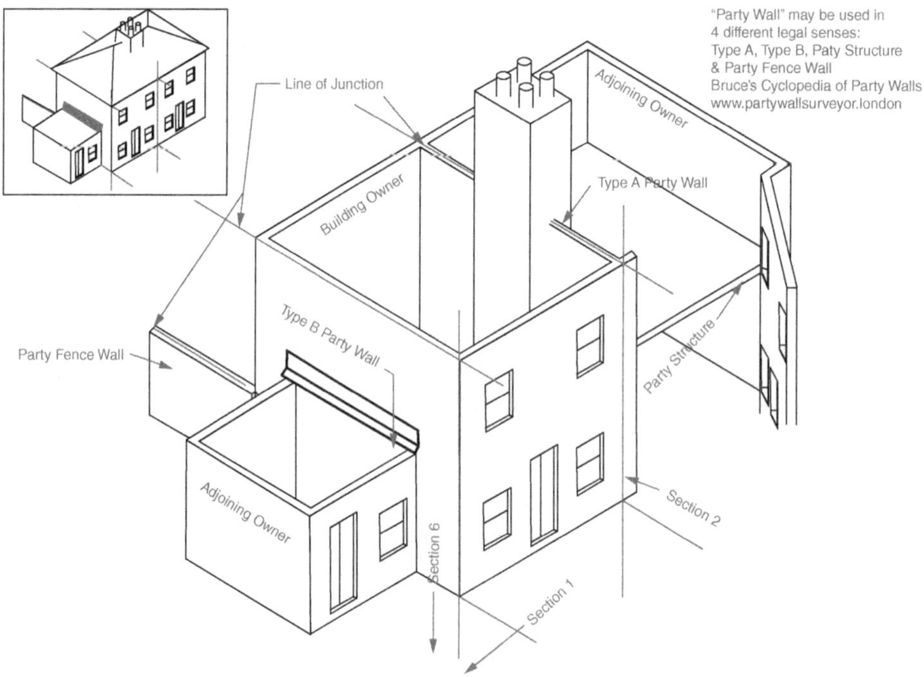

Diagram 2.1 Example of types of party walls in a building.[25]

24 Harvey – Norman Architects, 'Party Wall Act of 1996 – What is it and what are your obligations under it?', https://www.harveynormanarchitects.co.uk/articles/the-party-wall-act-of-1996-definitions-and-obligations
25 You can see the original coloured image in relation to Diagram 2.1 here: https://pwbsl.com/party-wall-surveyor/party-wall-surveyor/.

Diagram 2.1 sets out an illustration of the different types of party walls inside a building. Boundary walls are not party walls.[26] According to *Keating on Construction Contracts*:[27]

> *Boundary walls are referred to in the Act although not defined. It appears from the slightly complicated explanations in ss.1(1)(b) and 2(1) that a boundary wall can be either a party fence wall, the external wall of a building (both mentioned in those subsections), or a freestanding wall which only stands on the land of one party.*

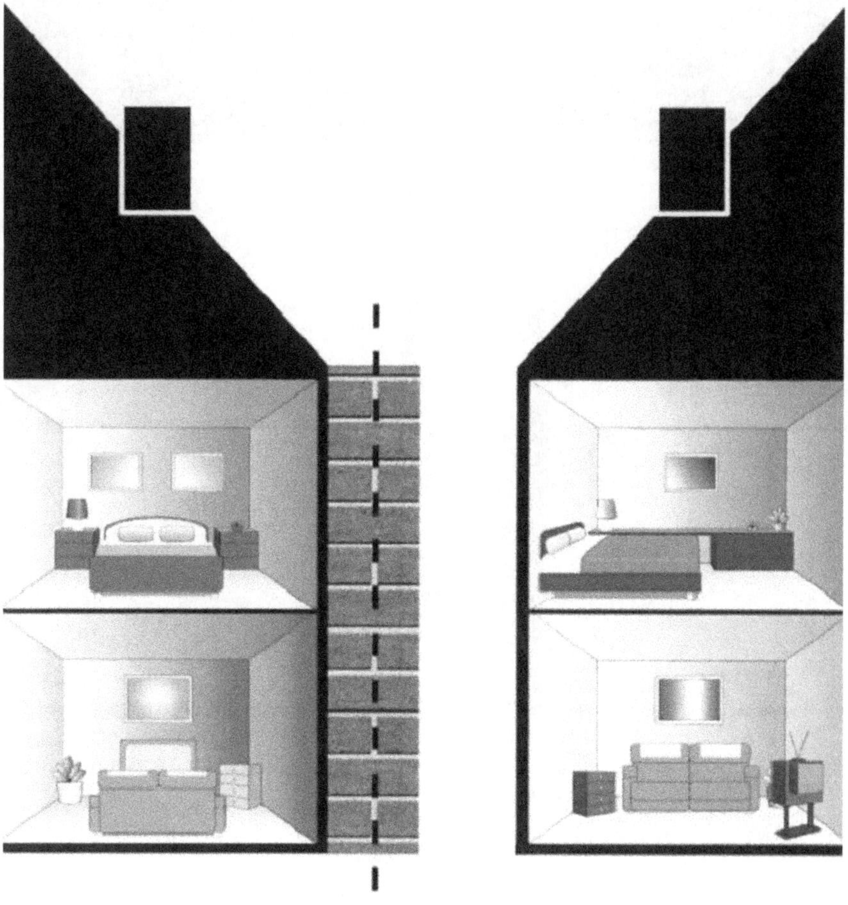

Diagram 2.2 Type A Wall forming part of one building.[28]

26 DCLG's Explanatory Booklet (n 26) 5.

27 (n 14) 16–089.

28 DCLG's Explanatory Booklet (n 26) 3. You can see the original coloured images in relation to Diagrams 2.2–2.9 here: https://www.gov.uk/government/publications/preventing-and-resolving-disputes-in-relation-to-party-walls/the-party-wall-etc-act-1996-explanatory-booklet.

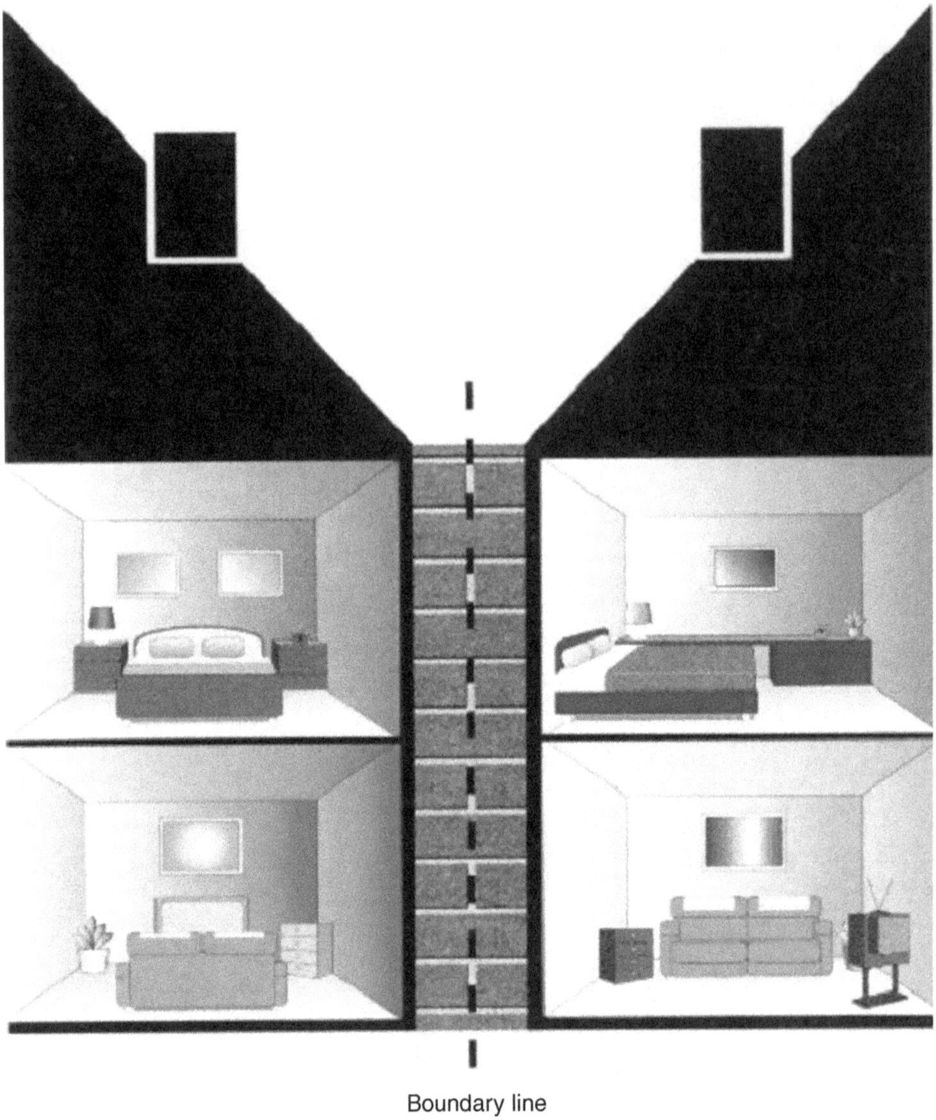

Diagram 2.3 Type A Wall forming part of two buildings.[29]

29 ibid. 3.

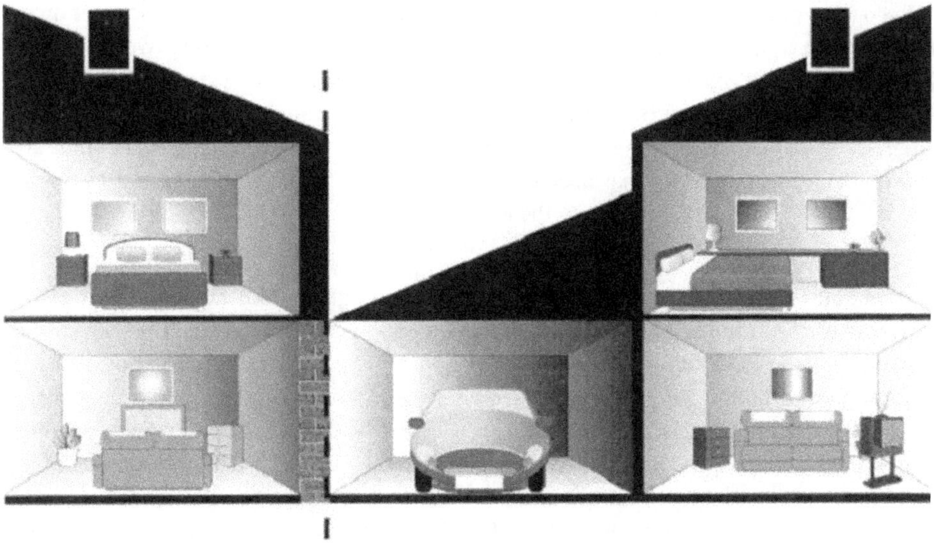

Diagram 2.4 Type B Wall.[30]

As shown in Diagram 2.4, there will be situations where a Type B Wall is longer or taller than the enclosed part of the adjoining owner's structure. In such a case, it is only the part of the wall which is enclosed to the neighbouring building that is an actual party wall. Therefore, in Diagram 2.4, the section of the wall which is above the garage (on the left) does not form part of the party wall.

Section 20(a) of the PWA 1996 is silent on the status of overhangs. By way of example, a wall can stand on one side of the boundary at ground level in its entirety. However, at the same time, it can be corbelled out at a level that is higher up, thus resulting in it overhanging the adjoining land. The question arises whether such a wall, or its upper part that is thickened, should be viewed as standing on lands of different owners. If so, it would mean that the word 'on' would include the word 'over' in this particular context. However, the answer seems to be 'no', which is supported by Section 2(2)(h) of the PWA 1996 as it specifically addresses overhangs, although there seems to be room for interpretation that such an overhanging wall is a party wall to the extent that it overhangs the adjoining land. One has to keep in mind the words of Jessel M.R. (albeit on a different type of case) on the approach to interpreting legislation, i.e. that courts have the duty to read legislation in a reasonable manner.[31] The wording of Section 20 of the PWA 1996 does not seem to contradict the interpretation, and a contradictory conclusion is noticeably inconvenient.[32]

30 ibid. 8.

31 *Standard Bank of British South America v Stokes* (n 5).

32 Bickford-Smith, Nicholls and Smith (n 22) 2.4.

The basic principle of land ownership and the tort of trespass is that a landowner owns their land up to the sky and down to the very depths. Where a building overhangs the neighbour's land, it constitutes a trespass. However, there may be an express easement, an estoppel or rights obtained by adverse possession entitling the owner to keep their overhang. Section 2 of the PWA 1996 gives rights to building owners where 'the said lands are built on or a boundary wall ... has been erected'. Therefore, there must be a building on the land before any rights are available to the building owner.

There are then two possibilities:

First, a wall on the ground that is not constructed straight up but at some point above ground level protrudes over the adjoining owner's land is a party wall all the way up and is to be treated as any other party wall for the purposes of a party structure notice.

Secondly, a wall/building on the adjoining owner's land overhanging the building owner's land or overhanging a party wall built at/astride the boundary (see the express wording of s 2(2)(h)). Such a second wall is not a party wall. However, it will interfere with any attempt by the building owner to build a vertical wall constituting a party wall (whether a new wall or raising an existing wall). This is where Section 2(2)(h) of the PWA 1996 becomes relevant, as it gives the building owner the right to demolish any part of the overhang so as "*to enable a vertical wall to be erected or raised against the wall or building of the adjoining owner.*" This right cannot be exercised so as to interfere with any easement the adjoining owner may have to maintain the overhang (Section 9(a) of the PWA 1996), but it will presumably override an acquired right through adverse possession or even an estoppel which falls short of an easement. However, the right is available only to enable the relevant vertical wall to be constructed. Merely building up to but remaining under the overhang does not engage the right. This second wall/building cannot be a party wall, as a wall which is not a party wall at its base on the ground is not capable of becoming a party wall higher up because it has an overhang.

Another concern is that Section 20(b) of the PWA 1996 raises the question as to how the PWA 1996 applies where a building encroaches illegally on adjacent land. Where a boundary wall is positioned on A's land entirely and B builds a building enclosing on it without permission, such a wall becomes a party wall falling under Section 20(b) of the PWA 1996. In terms of adverse possession, by having enclosed on A's wall, B is not only trespassing but is actually taking possession of its surface – that surface being part of A's land.[33] Where B serves notice trying to claim rights under the PWA 1996, A has the power to take action for removal of the trespassing building and is likely to succeed subject to B not having acquired title by adverse possession.[34] For B to be able to acquire immediate rights, it has to negotiate with A for these. For A to be able to keep full control of its wall, it should grant only a revocable licence to enclose on A's building. This will ensure that B's possession is prevented from being an adverse one.[35] This scenario seems to cover the options in terms of trespassing in relation to Section 20(b) of the PWA 1996.[36]

---

33 *Prudential Insurance Lt v Waterloo Real Estate Inc* [1999] EWCA Civ J0122-13.

34 Generally, 12 years adverse possession is required under the Land Registration Act 2002, para 1, Sch 6; *R. (on the application of Best) v Chief Land Registrar* [2015] EWCA Civ 17.

35 *Heslop v Burns* [1974] 1 W.L.R. 1241.

36 Bickford-Smith, Nicholls and Smith (n 22) 2.5.

2.3 Party Fence Wall

Another type of wall caught by the PWA 1996 is the party fence wall. Under Section 20 of the PWA 1996, this is a wall, which is not part of a building, standing on lands of different owners (astride the boundary between two properties) and is used or constructed to be used for separation of the adjoining lands:

> *... a wall (not being part of a building) which stands on lands of different owners and is used or constructed to be used for separating adjoining lands, but does not include a wall constructed on the land of one owner the artificially formed support of which projects into the land of another ...*

However, if the artificially formed support of a wall constructed on the land of one owner projects into the land of another owner, then it is not a party fence wall. What this refers to is a footing or foundation where Section 1(5) of the PWA 1996 applies.[37]

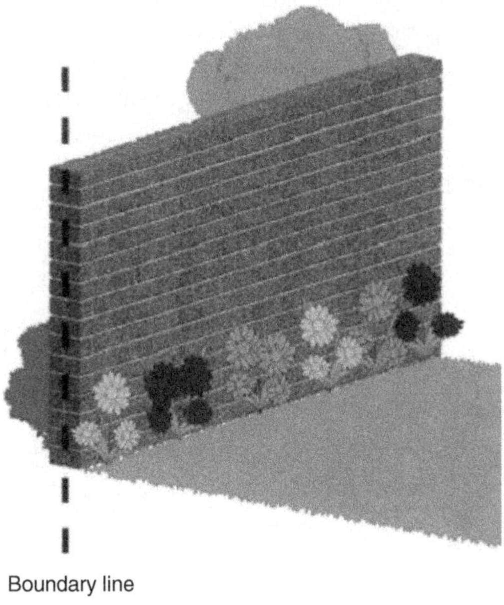

Diagram 2.5 Party fence wall (Part 1).[38]

Diagram 2.6 Party fence wall (Part 2).[39]

A party fence wall (as shown in Diagrams 2.5 and 2.6) could be, for example, a masonry garden wall.

For completeness, when it comes to party fence walls, one should bear in mind that if the wall encloses a railway embankment and the building owner wants to maintain or repair it, the building owner has no legal right of access to enter the railway. In such a case, the building owner will need to contact the relevant train company to obtain consent to do so, keeping in mind that this will be possible only if the services and the safety of the builders are maintained. This is because entering railway land without such consent is a

37 Isaac (n 23) 2–36.

38 DCLG's Explanatory Booklet (n 26) 3.

39 ibid.

criminal offence,[40] which is the case even for the purposes of repairing a structure by the building owner on the building owner's land. From a practical perspective, the relevant train company could do the repair works itself at its own cost and will be more inclined to do so if the building owner makes it clear to them that it is the relevant train company that is liable for actionable nuisance caused by its land damaging the building owner's wall.[41]

2.4 Party Structures

Finally, the PWA 1996 refers to party structures. According to Section 20 of the PWA 1996, a party structure is:

> *a party wall and also a floor partition or other structure separating buildings or parts of buildings, approached solely by separate staircases or separate entrances.*

All party walls are party structures, and the term also refers to horizontal party structures such as floors or ceilings between, for example, two adjoining flats or maisonettes as shown in Diagram 2.7. Roofs are not viewed usually as party structures. However, if the structure supporting it is shared between two properties, then that part is likely to be a party structure. Similarly, chimney stacks, which are shared between two properties, are also likely to be party walls or party structures.

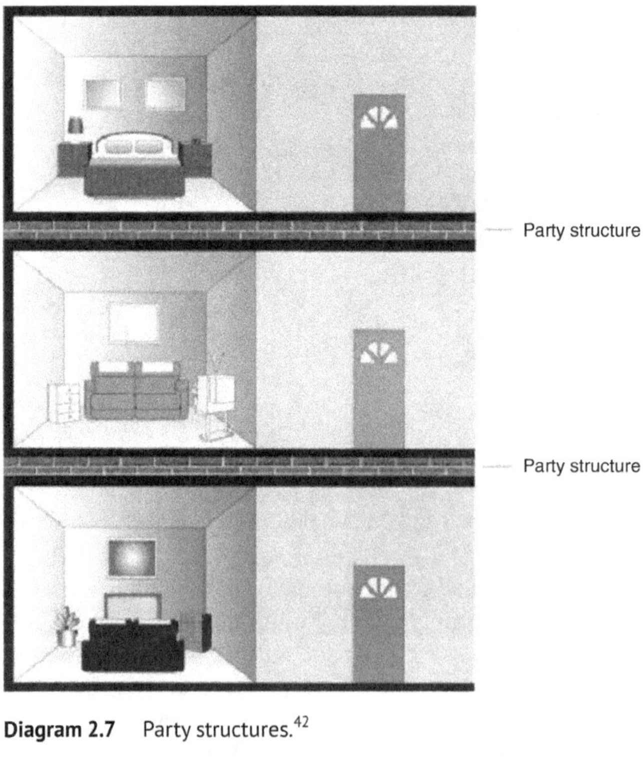

Diagram 2.7 Party structures.[42]

40 British Transport Commission Act 1945, s 55.
41 M Rousell, 'Is a wall enclosing a railway embankment a party wall?' *Practical Law UK, Ask*, Resource ID 5-522-4375 (8 July 2015), https://uk.practicallaw.thomsonreuters.com/5-522-4375?originationContext=document&transitionType=DocumentItem&contextData=%28sc.Default%29&comp=pluk.
42 DCLG's Explanatory Booklet (n 26) 3.

2.5 Excavations

The PWA 1996 also refers to certain types of excavation works in the context of party walls, as such excavations can have an effect on a party wall and related legal rights and obligations of the building owner and adjoining owner(s). In the case where the building owner wants to excavate near neighbouring buildings, it should inform the adjoining owner(s) by serving the relevant notice. Section 6 of the PWA 1996 provides for two options with regard to excavations:

a) works to excavate; and
b) works to excavate for and erect a building or structure within either:
   i) three metres from the adjoining owner's structure, if the proposed excavation would be at a lower level than the level of the bottom of the adjoining owner's foundations (Diagram 2.8); and
   ii) six metres from the adjoining owner's structure if any part of the proposed structures (typically foundations) would be dissected by a line drawn downwards at a 45-degrees angle from the nearest part of the adjoining owner's foundations (Diagram 2.9).

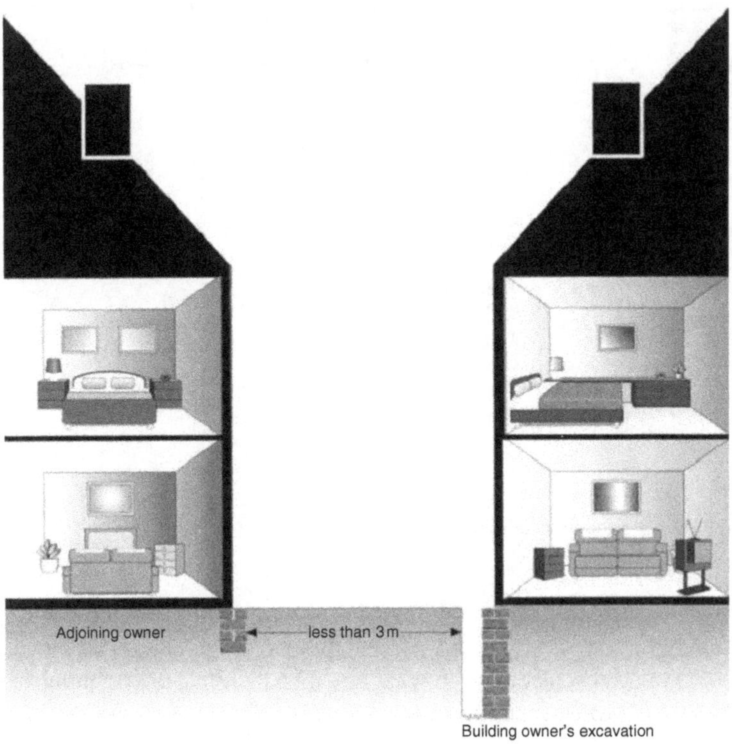

Diagram 2.8 Excavation – three metres from adjoining owner's structure.[43]

43 DCLG's Explanatory Booklet (n 26) 28.

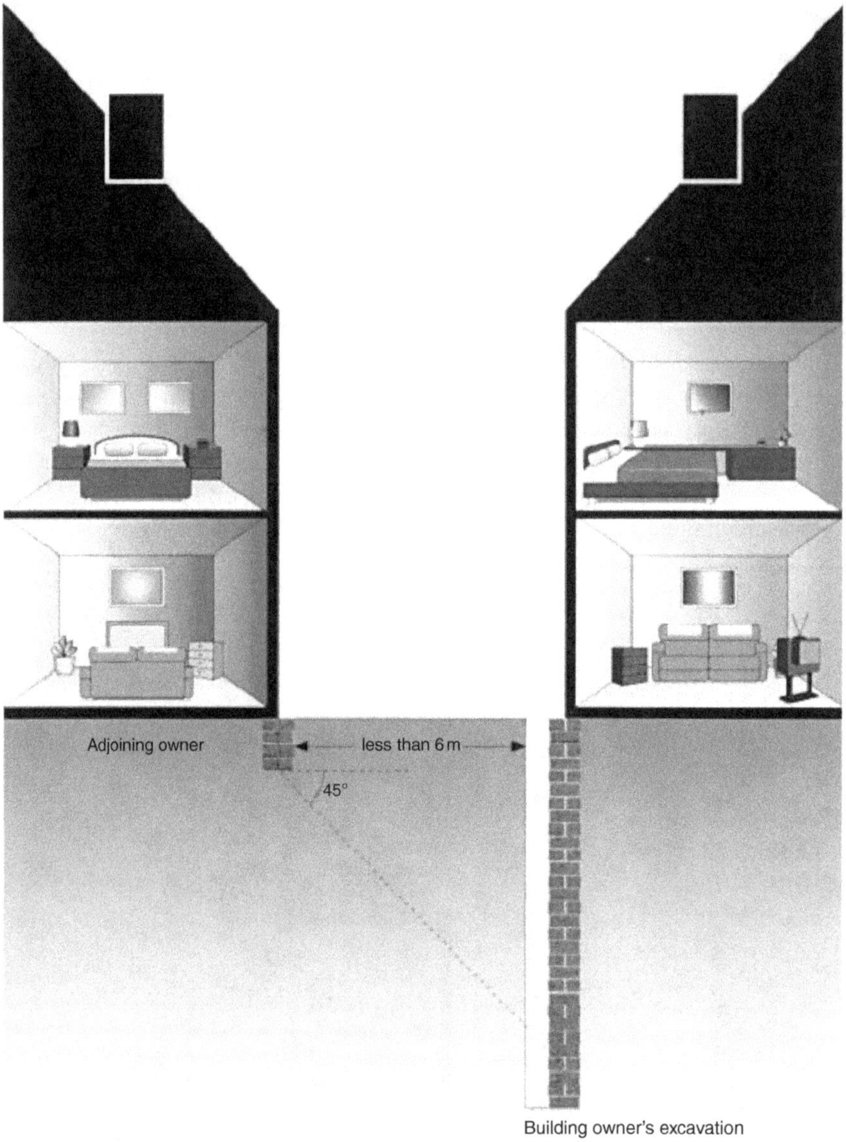

Diagram 2.9 Excavation – six metres from adjoining owner's structure.[44]

Liability for damage caused by tree roots and branches can cause private nuisance to the land of the Adjoining Owner. As the terminology 'building or structure' is not defined in the PWA 1996, there is some speculation as to what exactly a 'building or structure' is. It could be "... *anything which is constructed; and it involves the notion of something which is put together, consisting of a number of different things which are so put together or built*

44 ibid. 28.

together, constructed as to make one whole, which then is called a structure ..."[45] or even include a wall.[46]

2.6 Special Foundations

The PWA 1996 also refers to special foundations where they affect the rights and obligations of the building and adjoining owner(s) in the context of party walls. According to Section 20 of the PWA 1996, special foundations relate to a: "... *foundation in which an assemblage of beams or rods is employed for the purpose of distributing any load ...*". For historical context, the position adopted by Section 45(2) of the London Building Acts (Amendment) Act 1939 changed the words 'grillage foundations' to 'special foundations', which is the same position adopted by the PWA 1996.

According to Section 7 of the PWA 1996, special foundations on the adjoining owner's land require the adjoining owner's previous (i.e. prior to the start of the works) consent in writing (verbal consent by the adjoining owner is not sufficient). The adjoining owner can give a counter-notice thereby making its consent conditional. However, the matter is not as straightforward. The use of special foundations, together with their sections and details of construction, must be included in the building owner's party structure notice under Section 3(1)(b) of the PWA 1996. The adjoining owner then has a choice:

First, the adjoining owner can refuse the use of special foundations and that is the end of the matter.

Secondly, the adjoining owner may serve a counter-notice. However, there is a catch in the PWA 1996. An adjoining owner served with a party structure notice giving details of proposed special foundations "*to which the adjoining owner consents*" may serve a counter-notice under Section 4(1)(b) of the PWA 1996 requiring the special foundations to be placed at a specified greater depth or of sufficient strength to bear the load to be carried by columns of any intended building of the adjoining owner. The terms of this subsection are important. The adjoining owner must first consent to the special foundations under Section 7(4) of the PWA 1996. Only then the adjoining owner may require either or both of the criteria of a) "*a greater specified depth*"; or b) "*be constructed of sufficient strength to bear the load to be carried by columns of any intended building of the adjoining owner*". The issue for the adjoining owner is that it has already consented to the special foundations, and although Section 4(1)(b) of the PWA 1996 refers to a 'requirement', the building owner might be able to get out of compliance under Section 4(3) of the PWA 1996.

The adjoining owner will have to bear the costs of the additional works specified in its counter notice (Section 11(9) of the PWA 1996). Also note Section 11(10) of the PWA 1996, which relates specifically to special foundations. While the definition was originally aimed at grillage foundations, it can also include reinforced concrete foundations.

45 *Hobday v Nicholl* [1944] 1 All E.R. 302.
46 *Mills & Rockleys Ltd v Leicester City Council* [1946] K.B. 315.

2.7 Conclusion

The definitions and clarifications as to what party walls are/can be and what works affect them lay the ground for understanding when and how certain rights and obligations of building owners and adjoining owners arise and what relevant procedures apply to the parties under the PWA 1996. The potential for disputes can arise when properties are renovated, changed or when new properties are built. This is partly the reason why the PWA 1996 has been introduced (elaborated on in Chapter III) to assist with regulating alterations to buildings, land, fences and walls. While the PWA 1996 is helpful, it does not cover all areas of law and fact that can affect party walls and linked neighbourly relationships between building owners and adjoining owners.

3

Statutory Rights – Party Wall etc. Act 1996

3.1 History behind the Party Wall etc. Act 1996

Now that the meaning of the term 'party wall' has been clarified (as well as the terminology linked to works related to party walls and relevant parties), let us examine the historical evolution of the legislation behind the PWA 1996.

The area of party wall matters has been in the making for centuries. Before the PWA 1996, there were a number of different principles that applied to party walls throughout England and Wales. London, for example, was governed by a number of Building Acts with Pt. VI of the London Building Acts (Amendment) Act 1939 being the latest. Then there were other parts of the country, including Bristol, which were governed by local acts. According to *Keating on Construction Contracts*:[47]

> *Prior to the Party Wall, etc. Act 1996, there were a number of different principles applicable to party walls across England and Wales. London was governed by a series of Building Acts, the latest being Pt VI of the London Building Acts (Amendment) Act 1939 ("the 1939 Act"). Other parts of the country (including Bristol) were governed by local acts. The majority of the country, however, was not governed by statute at all but by a complex system of common law dependent on the type of party wall under consideration. The Party Wall, etc. Act came into force on 1 July 1997. The 1939 Act and other local Acts were repealed at the same time. The whole of England and Wales is now governed by the statutory scheme in the Party Wall, etc. Act, which is very similar to the previous London scheme. The purpose of the Act is twofold. First, it allows an owner (the "Building Owner") to carry out works to a party structure or to make use of a party structure. Without the sanction of the Act, some of these works would constitute a nuisance and/or a trespass. Secondly, it provides safeguards to the "Adjoining Owner" where works are carried out to a party structure. Prior to the Act, an adjoining owner could have been left without a remedy, for example, in a situation where the removal of the building*

47 (n 14) 16-083.

Party Wall Disputes: The Party Wall etc. Act 1996 and Beyond – *Legal Coherence and Dispute Management*, First Edition. Dr Laura Lintott.

owner's half of the party wall left their half exposed to the weather. Safeguards provided include requirements to give notice to "Adjoining Owners" prior to carrying out any works and to make good any damage caused.

The thrust of the PWA 1996 is that it grants rights which are inconsistent with and thus override the common law. Section 2 gives the building owner rights which it does not have in common law and doing so overrides the law of trespass. Section 7(1) provides that the building owner can carry out its works under these rights free from an action in nuisance provided it does not cause 'unnecessary inconvenience' to any adjoining owner or occupier. How far 'unnecessary inconvenience' differs from the tort of nuisance may be debated. In return, the adjoining owner is given various rights (see *inter alia* sections 2(3)–(7), 4, 7(2), 7(3) or 12). The right to compensation directly against the building owner is particularly useful, as it avoids an adjoining owner having to sue the builder (who will almost certainly be an independent contractor) and may not be insured. There are also important rights arising under section 11 (Expenses).

The roots of party wall legislation in London date back to Henry Fitz-Alwyn's assize of 1189. The initial motivation behind the legislation was to prevent the spread of fire by requiring substantial stone walls to be constructed between houses. However, as the Great Fire of London of 1666 shows, enforcing this rule was not as effective as it should have been. The Fires Prevention (Metropolis) Act 1774 (**FPMA 1774**)[48] covered party walls extensively and gave owners the power to raise party walls, as long as these were sufficiently thick and in compliance with the FPMA 1774. It also contained provisions for the appointment of statutory surveyors. The oldest piece of legislation covering party walls resembling the modern form is Part III of the Metropolitan Building Act 1855.[49]

The next piece of legislation in the evolutionary process of party walls regulation was the London Building Act 1894 (amended in 1905).[50] This was then superseded by the London Building Act 1930,[51] but this step was not considered entirely satisfactory, and several provisions had been superseded due to changes in construction technology. As a result, relevant provisions of the London Building Act 1930 were re-enacted with amendments in amendments in Part VI of the London Building Acts (Amendment) Act 1939.[52] It was to be construed as one with the London Building Act 1930 (together, the **1939 Act**). The 1939 Act applied only to the Inner London Boroughs.[53] Interestingly, statutes that adopted the wording of the English pieces of legislation back then were passed in a number of states of Australia in the 19th century in order to control construction work in larger cities.[54]

48 14 Geo 3 c LXXVIII.
49 18 & 19 Vict c CXXII.
50 57 & 58 Vict. c CCXIII and 5 Edw. VII c CCIX, respectively.
51 20 & 21 Geo 5 c CIVIII.
52 2 & 3 Geo 6 c XCVII.
53 London Government Act 1963, section 43.
54 Examples include the Melbourne Building Act 1849 (13 Vict No 39) (Vic), Launceston Building Acts 1854 and 1894 (18 Vict No 18, 58 Vict No 32), (Tas), Sydney Building Act 1879 (42 Vict No 25) (NSW), Hobart Building Act 1918 (9 Geo V No 75) (Tas).

The PWA 1996[55] was introduced by the Earl of Lytton (a practising chartered surveyor) as a Private Members Bill to the House of Lords, received Royal Assent on 18 July 1996 and came into force on 1 July 1997 by the Party Wall etc. Act 1996 (Commencement) Order 1997.[56]

According to the Earl of Lytton, as he stated in the House of Lords debate on 31 January 1996:[57]

> *The aims of the Bill are to extend the tried and tested provisions of the London Building Acts to England and Wales. It rests upon a principle of voluntary agreement between parties wherever possible; it provides for notice to be given where works are proposed; there is an opportunity to respond and comment; it sets out to protect existing structures; there is a clear liability for damage and making good; there is provision for the resolution of disputes, other than by going to law; it sets out how costs of works and fees arising from them shall be dealt with; and clarifies the extent of rights over common structures, including floors – that is, floors between different units of occupation.*

The Earl of Lytton also noted that, at the time, the Bill is a "*safety net and not a fiery hoop*," which makes sense as, while a building owner's failure to trigger the procedure set out in it amounts to a statutory breach of the PWA 1996, this does not result in criminal liability. Another point the Earl of Lytton raised in the same debate was that the Bill included changes to adapt local legislation to national purposes and that one of the reasons to pass the Bill included the need to rectify an "*awkward legal precedent to do with party walls used by only one owner, and it sets out to remedy an anomaly relating to different treatment of floor structures.*"[58]

One of the criticisms raised (by the Earl of Kinnoull) in the debate was the Bill's definition of 'surveyor'.

> *There are two matters which concern me in that context. First, there is the definition of "surveyor" in Clause 18. I fully understand and accept that the word "surveyor" is very much a generic term, as the noble Earl said. I also fully understand that it does not have the meaning of surveyor as we know it – a chartered surveyor, or whatever. Basically, it means a competent or a relevant professional person. However, it does not say that in the Bill. I find that very surprising when we think of the seriousness of the subject. The Bill simply refers to a "person." That person could be either competent or incompetent.*

There is no regulation or safeguard of the quality of surveyors to deal with party wall disputes. The issue of not having any safeguards in place to ensure quality and consistency of surveyors and their party wall awards persists in the current PWA 1996.

---

55 Party Wall etc. Act 1996, 'Introduction'.

56 SI 1997/670.

57 Vol 568 cc1535–48, https://api.parliament.uk/historic-hansard/lords/1996/jan/31/party-wall-bill-hl.

58 House of Lords Debate, 31 January 1996, Vol 568 cc1537, https://api.parliament.uk/historic-hansard/lords/1996/jan/31/party-wall-bill-hl.

The general view in the debate of the Lords at the time was a surprise over having party wall legislation (prior to the Bill) in place that does not consistently apply to areas outside of London.

The debate continued on 22 May 1996,[59] where the Earl of Lytton incorporated a number of amendments to the Bill having taken into consideration the previous debate on 31 January 1996. The Lords' overwhelming support in favour of passing the PWA 1996 seems to be largely due to the recognition of the fact that party wall disputes are highly contentious in nature and, at the time, the legislation on party walls did not apply in a consistent manner to England and Wales.

Section 21(1) of the PWA 1996 confers the power on the Secretary of State to amend/repeal earlier local Acts. This allowed Part VI of the 1939 Act and sections 27 and 32 of the Bristol Improvement Act 1847 to be repealed.[60] The Earl of Lytton confirmed that the main reason behind enacting the PWA 1996 was to ensure that party wall legislation essentially extended the provisions of the 1939 Act to the entirety of England and Wales.[61]

Local Acts that contain provisions that address party wall issues other than those that have been specifically repealed are not affected by the PWA 1996 and the commencement or repeal orders.[62]

While the 1939 Act was repealed by the PWA 1996, case law on the 1939 Act and further legislative predecessors are still of relevance and aid interpretation of the PWA 1996 provisions. This follows the principle of interpretation from the case *Barras v Aberdeen Steam Trawling & Fishing Co Ltd:*[63]

> *It has long been a well established principle to be applied in the consideration of Acts of Parliament that where a word of doubtful meaning has received a clear judicial interpretation, the subsequent statute which incorporates the same word or the same phrase in a similar context, must be construed so that the word or phrase is interpreted according to the meaning that has previously been assigned to it.*

Where there is case law on previous versions of the PWA 1996, such case law is likely to be followed when it comes to the issue how the PWA 1996 should be interpreted.[64] This is also why this book refers to numerous cases pre-dating the PWA 1996 as these cases are well placed in aiding interpretation of the PWA 1996 as well as covering gaps the PWA 1996 does not cover when it comes to other areas of law and fact.

59 House of Lords Debate, 22 May 1996, Vol 572 cc931–55, https://api.parliament.uk/historic-hansard/lords/1996/may/22/party-wall-bill-hl.
60 Bickford-Smith, Nicholls and Smith (n 22) 1.1–1.6; 1.25–1.27.
61 *Hansard* HL Debates Vol 568, col 1536 (31 January 1996); confirmed in *Kaye v Lawrence* [2010] EWHC 2678 (TCC).
62 Bickford-Smith, Nicholls and Smith (n 22) 1.27.
63 [1933] AC 402, 411.
64 Bickford-Smith, Nicholls and Smith (n 22) 1.6.

3.2 The Purposes of and Main Areas Covered by the Party Wall etc. Act 1996

The PWA 1996 mainly emerged due to the fact that construction works by a building owner can result in damage being sustained to structures or buildings of an adjoining owner and/or affect the adjoining owner's use or enjoyment of the structure or party wall.

Stephen Bickford-Smith concisely summarises the operation of the PWA 1996:[65]

> *Party walls separating the lands of adjoining owners are sensitive areas at common law. Since each owner has some interest in the wall, often neither can do any work to it without the consent of the other, which may lead to paralysis. The broad object of the Act is to set up machinery enabling the building owner to carry out works within the scope of the Act. The machinery starts with a notice under the Act, and, unless the adjoining owner consents, this sets in train an elaborate procedure for referring the matter to surveyors for determination. The surveyors embody their decision in an award which, subject to the possibility of appeal to the county court, is binding on both parties.*

There are two main purposes the PWA 1996 has. First, it makes it possible for the building owner to carry out works to a party structure or to make use of it. This is because, had the PWA 1996 not been in place, some of these works would amount to nuisance and/or trespass. The second purpose of the PWA 1996 is that it gives safeguards to adjoining owners where works are carried out to a party structure. Again, had the PWA 1996 not been in place, adjoining owners could have ended up in situations without having a straightforward recourse to remedy. This would be the case, for example, where the building owner removes its half of the party wall, which causes the adjoining owner's half to be exposed to the weather. The PWA 1996 provides for safeguards, which include the need to give notice to adjoining owners before building owners carry out any relevant works and ensuring that they make good any damage caused by such works.[66]

The PWA 1996 attempts to give a structure to property neighbours in England and Wales who share a party wall enabling them to carry out related construction work falling under the PWA 1996. It is concerned with a) building a new party wall or a wall adjacent to its side of the boundary;[67] b) carrying out relevant works;[68] c) excavating, or excavating for and erecting a building or structure within three or six metres from the building or structure of an adjoining owner;[69] and d) construction of special foundations or placing foundations on the adjoining owner's land.[70]

---

65 ibid. 1.7.

66 *Keating on Construction Contracts* (n 14) 16-083.

67 PWA 1996, s 1.

68 ibid., s 2.

69 ibid., s 6.

70 ibid., ss 7, 20.

3.3 Areas of Uncertainty

However, it should be borne in mind that there are exceptions to the PWA 1996 applying to England and Wales, and one should check in each case whether such exceptions apply or not. For example, one of such exceptions is Inner London that belongs to the Inns of Court.[71] Additionally, for example, the PWA 1996 does not apply to specific works that fall under legislation applying to the Crossrail railway construction in London.[72]

As noted earlier, the PWA 1996 refers to adjoining owners and building owners. Under section 20 of the PWA 1996, 'adjoining owner' or 'Adjoining Occupier' means "*any owner and any occupier of land, buildings, storeys or rooms adjoining those of the buildings of the building owner and for the purposes only of section 6 within the distances specified in that section.*"

3.3.1 Scale

There are works that are considered to be too small to be caught by the PWA 1996, which include: drilling into a party wall to fix plugs and screws for ordinary wall units or shelving; cutting into a party wall to add or replace recessed electric wiring and socket and removing old plaster and re-plastering. The distinguishing feature is whether the work has any consequences for the structural strength and support of the party wall as a whole or could cause damage to the adjoining owner's side of the party wall.[73]

It very much depends on the individual circumstances of each case although, in general and as mentioned above, removing old plaster would not be considered to fall under the PWA 1996. One could also argue that the point is (rather than scale) that the works listed at a)–d) above are restricted to the owner's half of the wall (divided vertically).

These nuances are further aided by case law. In terms of the issue of plaster removal, in *Kelliher v Ash Estates Holding Ltd and another,*[74] it was held that section 2(2)(f) of the PWA 1996 could apply in relation to this. It depended on the way in which the work was done. If the plaster was in poor condition and could be removed easily, it would not fall within the scope of the PWA 1996. If an electric tool was needed to remove it, which could go into the edge of the brickwork, that would be substantial cutting into the party structure and hence would be caught by the PWA 1996. This is an example of how important it is to view legislation in the context of case law, which can aid in how legislation could or should be interpreted where the answer is not immediately obvious from legislation or where there are gaps.

3.3.2 Location

Another slightly unclear area are chimney breasts removed by a party on its side of a party wall. As per section 2(2)(g) of the PWA 1996, if the works include cutting away from a party wall any projecting chimney breast, this would be caught by the PWA 1996.

---

71 ibid., s 18.
72 Crossrail Act 2008, Schedule 14, p 17.
73 DCLG's Explanatory Booklet (n 26) 6.
74 (CLCC, 15 July 2013).

3.3.3 What Does the Party Wall etc. Act 1996 Not Do?

The objective of the PWA 1996 is to prevent disputes from arising between neighbours by allowing for certain works to be done which would otherwise be treated as nuisance or trespass. Alternatively, the objective of the PWA 1996 is to help parties to party wall disputes to navigate through them. It should be noted that complying with the PWA 1996 does not absolve the parties from their duties with regard to obtaining planning permissions and/or complying with building regulations (which are both covered by separate legislation) or other legislation.

According to *Nicholas Isaac*:[75]

> *The Act, like its predecessors, operates to vary the common law as it relates to the difficult and potentially highly contentious physical and legal area where two people's ownership of land meets.*

The question is what the PWA 1996 does not do. According to Stephen Bickford-Smith:[76]

> *The Act does not make it a criminal offence not to comply with its provisions (except failure to allow entry to premises). Nor does it provide any supervisory role by the local council or government to see that individuals comply with the Act.*

As noted above, the notice provision mechanism (further elaborated on in Chapter IV) set out in the PWA 1996 does not criminalise the building owner's failure to trigger it.

3.3.4 'Construction Rights'

To take it further, the PWA 1996 does not fully cover the different areas of law and fact discussed in this book. A 'construction right' could be described as a right to carry out construction work which is not illegal or tortious or in breach of contract. Rights under section 2(2) of the PWA 1996 relate to rights to carry out work which would otherwise amount to tort. As noted earlier, an essential feature of the PWA 1996 is that it overrides the common law in certain specified circumstances.

Where an entire wall is in co-ownership, either owner may prevent the other from doing nearly anything to any part of the wall without consent. However, overwhelmingly, party walls are owned by the two owners on the basis of a longitudinal separation. Of course, either party has the right to carry out work which is restricted to its side of the wall (such as putting a screw or nail into a (party) wall on which to hang a picture or putting up shelves). The 'statutory rights' are rights which cover the entire wall and are important for that reason; they allow the building owner to override any objection the adjoining owner may have to interference with 'its side of the wall'. Common law and statutory rights do co-exist, with the PWA 1996 rights being rights of a different nature to the rights arising in common law and overriding common law rights where they are inconsistent.

75 Isaac (n 23) 1–25.

76 Bickford-Smith and Lamont, 'Party Wall etc. Act 1996 – 10 years on' (n 17) 2.

3.3.5 Potential for Reforms

The PWA 1996 has been around for over quarter of a century at the time of publication of this book. While the PWA 1996 was a positive step forward in relation to party wall disputes in particular, legislative and policy improvements should continue to be made to save public money. The PWA 1996 was drafted at pace and little time was given to improve upon some inconsistencies borrowed from its legislative predecessors.

It would be helpful to introduce or update certain definitions in the PWA 1996, for example, formulating a definition of 'party wall surveyor', as regulation of party wall surveyors (ideally qualified professionals who are subject to professional body disciplinary procedures) would lead to an improvement of the quality of party wall awards. Another improvement would be to make it clear what rights, if any, pass to successors in title. Provision for the Secretary of State, or Lord Chancellor, to approve a procedural code for party wall surveyors to follow when making awards would seem sensible.

It would also be useful to impose more severe consequences for a building owner who fails to issue the relevant notice under the PWA 1996 while also giving a limited opportunity to the building owner to fix absence of such notice to enable the PWA 1996 to apply and for the adjoining owner to avoid the necessity of going to court.

As time has passed, the need for reforms has become clear to ensure that members of society do not waste time, cost, emotion and energy unnecessarily on disputes that should be better structured. Both statutory and policy reforms are needed to ensure that courts are not burdened by party wall disputes unnecessarily, that ADR (such as adjudication, mediation, expert determination or even arbitration) is mandatorily considered and genuinely attempted by parties to party wall disputes. The rationale is similar to that behind the Housing Grants, Construction and Regeneration Act 1996 (as amended), which is very popular in the construction industry and the Pre-Action Protocol for Construction and Engineering Disputes (2nd edition).

The creation of a party wall disputes area of its own merit should lead to regulating and monitoring the quality of party wall surveyors and their awards, to ensure justice is served correctly, and preventing unnecessary hardship to disputants.

However, such new legislation relating to party walls should go further to provide a mechanism or tool to solve all types of disputes that can relate to party walls (not only those currently covered by the PWA 1996). One option is to expand the existing statutory procedure under the PWA 1996. Another option is to replace the PWA 1996 altogether with new legislation setting out in greater detail what ADR mechanisms the parties should use, when, why and how.

The dispute resolution mechanism under the PWA 1996 relates only to disputes arising under the PWA 1996 and does not encompass all possible disputes related to party walls, such as those relating to easements, tort or boundary disputes, thus leaving questions within disputes unanswered. In connection with reforming/replacing the PWA 1996, as its provisions are currently limited and not able to cover many of the complexities linked to party wall disputes, a checklist could be put in place that members of the public should have regard to when involved in a party wall dispute. This checklist could be presented either in the same manner as the current DCLG Explanatory Booklet, alternatively as an annex to the Technology and Construction Court Guide, or as part of a Practice Direction (although such a Practice Direction would require input from the Rules Committee, and this may be too prescriptive). While an 'annex' or 'booklet' route would seem to be preferable (as these would serve as guidance rather than being prescriptive and would be more flexible to prepare and to revise), the Practice Direction route is still worth exploring.

In addition, there is the question of retrospective party wall awards. Currently, where the relevant notice has not been served under the PWA 1996 and works have started and/or have been completed without the service of such notice, the PWA 1996 does not apply retrospectively and therefore any related dispute would have to be decided in court, unless parties agree to follow the procedure under the PWA 1996 retrospectively or unless they agree to an ADR avenue. A solution to this issue would be to amend the PWA 1996 so that it applies retrospectively, or to ensure that any superseding legislation has a retrospective effect.

3.4 Rights and Obligations of the Parties Under the Party Wall etc. Act 1996[77]

The PWA 1996 confers extensive rights on those that carry out works falling under the PWA 1996. Such rights include the right to enter land, remove furniture/fittings or even break open doors or fences to enter premises if a police officer accompanies such a party. Conversely, the building owner is obliged to compensate the adjoining owner or Adjoining Occupier in relation to any loss/damage that has been caused by carrying out of the works. The PWA 1996 ensures that a building owner that failed to comply with the PWA 1996 when carrying out works that fall within the PWA 1996, is liable in nuisance or trespass. A mandatory injunction necessitating removal/reversal of unauthorised works may be granted by a court in exceptional situations.[78]

The rights and obligations conferred by the PWA 1996 and detailed below only apply if the building owner decides to follow the relevant notice mechanism under the PWA 1996.

The building owner has a number of statutory rights. Section 2 of the PWA 1996 sets out what work can be performed. Most commonly used rights are:[79]

a) party wall repairs;
b) insertion of a damp proof course;
c) underpinning the whole thickness of a party wall (for example, to prevent settlement or to form a basement);
d) cutting into a party wall to take the bearing of a beam (such as a loft conversion);
e) raising the height of a party wall (for example, adding a new storey);

---

77 T Reid, 'The Party Wall etc. Act 1996 (PWA 1996)' (Maintained) *Practical Law UK Practice Note*, Resource ID 8-383-5739, https://uk.practicallaw.thomsonreuters.com/Document/Ib5556c44e83211e398db8b09b4f043e0/View/FullText.html?transitionType=SearchItem&contextData=(sc.Search)#co_anchor_a952869.

78 Megarry & Wade (n 8) 29-051.

79 DCLG's Explanatory Booklet (n 26) 4.

f) extending a party wall downwards (such as forming a basement);[80]
g) demolishing and rebuilding a party wall (for example, if it is structurally defective); and
h) cutting off projections from a party wall (or from an adjoining owner's boundary or external wall) if necessary to build a new wall adjacent to that wall (for example, removing a chimney breast).

The building owner has a number of statutory obligations with regards to works relating to party walls and caught under the PWA 1996:

a) informing all adjoining owners prior to any intended works under section 2 of the PWA 1996;[81]
b) serving a notice on the adjoining owner, which must be clear enough so that the adjoining owner understands which counter-notice to give in response;[82]
c) exercising reasonable care when carrying out the works;[83]
d) not causing unnecessary inconvenience to the adjoining owner during the works under sections 4 and 7 of the PWA 1996;[84]
e) providing temporary protection for adjacent buildings and property where necessary;[85]
f) making good any damage caused by the works or paying in lieu of making good if the adjoining owner requests it under section 11(8) of the PWA 1996;[86]
g) commencing the works, with due diligence,[87] within 12 months of the date of the notice so that the building owner does not lose the rights under the notice;[88]
h) carrying out the works as agreed between the building owner and the adjoining owner or as set out in the party wall award under section 7(5) of the PWA 1996 if there is a dispute under section 10 of the PWA 1996;
i) where party walls and structures are modified, repaired or demolished and rebuilt under section 2(2)(a) and (b) of the PWA 1996, section 11(4) and (5) states that the cost of the work is to be shared where the work is necessary due to a defect or want of repair, in proportion to the use each of the parties makes of the structure or wall and the responsibility of each party for the defect or want of repair in question;[89] and

80 While this is listed in DCLG's Explanatory Booklet (n 26) 4, 'extending a party wall downwards' is a point of contention, particularly after the decision in *Chaturachinda v Fairholme* (CCCL, 23 September 2015), where HHJ Bailey discussed the meaning of 'special foundations'. The question is whether 'extending downwards' can or cannot be implied into the meaning of section 2(2)(a) of the PWA 1996, which confers rights to 'underpin, thicken or raise'. There is some debate on the meaning of 'extending downwards' particularly in the context of reinforced concrete underpinning and any potential need for consent under section 7(4) of the PWA 1996.
81 DCLG's Explanatory Booklet (n 26) 4.
82 *Hobbs Hart & Co v Grover* [1899] 1 Ch. 11.
83 *Alcock (Douglas Alexander) v Wraith* [1991] 12 WLUK 156.
84 DCLG's Explanatory Booklet (n 26) 5.
85 ibid.
86 ibid.
87 *West Faulkner Associates v Newham LBC* (1995) 11 Const. L.J. 157, 161:

What in particular the word 'diligently' contributes to the concept is the need to apply that physical capacity industriously and efficiently towards that same end.

Taken together the obligation upon the contractor is essentially to proceed continuously, industriously and efficiently with appropriate physical resources so as to progress the works steadily towards completion substantially in accordance with the contractual requirements as to time, sequence and quality of work.

88 PWA 1996, ss 1, 3, 6.
89 DCLG's Explanatory Booklet (n 26) 5.

j) if use is made of party walls that were previously built at the adjoining owner's cost, the PWA 1996 provides for a fair payment to be made to the adjoining owner.[90]

The adjoining owner's rights conferred by the PWA 1996 (as set out in DCLG's Explanatory Booklet)[91] include:

a) raising a dispute in response to the building owner's notice under section 5 of the PWA 1996 (arguably, this is a consequence of the deeming provision rather than a right);
b) by way of a counter-notice, requiring the building owner to add additional works into the building owner's notice under section 4 of the PWA 1996;
c) appointing a party wall surveyor to resolve any dispute under section 10 of the PWA 1996;
d) requiring reasonably necessary measures to be taken in order to protect the adjoining owner's property from foreseeable damage and for the adjoining owner's security, for example:
   A) not to be caused any unnecessary inconvenience under sections 4 and 7 of the PWA 1996 (to clarify the wording in DCLG's Explanatory Booklet, it is a building owner's duty to avoid unnecessary inconvenience rather than an adjoining owner' right); and
   B) asking for security for expenses under section 12 of the PWA 1996 before the building owner starts work under the PWA 1996 to guard against the risk of being left in difficulties if the building owner stops work at an inconvenient stage;
e) compensation for any loss or damage caused by relevant works under sections 1(7) and 7 of the PWA 1996.

The adjoining owner's statutory obligations under the PWA 1996 include those listed below.

a) allowing the building owner and its workmen access to the adjoining owner's land under relevant section 8 of the PWA 1996 after receiving the notice. Such right of entry also includes a surveyor appointed or selected under section 10 of the PWA 1996, erection of scaffolding[92] or removing any furniture or fittings by the building owner or his workmen;
b) serving a counter-notice within one month of service of the building owner's notice if the adjoining owner requires additional works to be incorporated into the building owner's works on an existing party wall (section 4 of the PWA 1996); and
c) contributing to the cost of the work if the adjoining owner requires the building owner to carry out additional works or agrees to the construction of a new party wall (section 11 of the PWA 1996).

According to HHJ Grant in *Heathcote v Doal*:[93]

> *... the whole point of the Party Wall Act is to provide for a regime of the service of notices and counter-notices, the appointment of surveyors, the provision of an award, all of which is intended to be done before works are commenced. ...*

---

90 ibid.
91 ibid. 21.
92 ibid. 25.
93 [2017] 1 WLUK 613, 40.

It means that the PWA 1996 aims to give clarity and a structured process to adhere to in order to avoid unnecessary confusion in terms of which party is responsible for which works, damage, cost and prevent conflict where possible. In addition, if a dispute arises anyway with regard to party walls, the PWA 1996 provides for how to proceed. However, the following chapters will also focus on areas that go beyond the realm of the PWA 1996, and Chapter IV discusses the dispute resolution process under the PWA 1996 as well as other dispute resolution avenues that are available to the relevant parties.

3.5 Party Wall etc. Act 1996 (Electronic Communication) Order 2016 (SI 2016/335)

Before moving on to separate pieces of legislation from the PWA 1996, the Party Wall etc. Act 1996 (Electronic Communication) Order 2016 (SI 2016/335) (**Order**) needs to be mentioned as it allows a notice or other document served under the PWA 1996 to be capable of being served on a person by way of an electronic communication sent to an electronic address that the recipient has specified and if such a recipient has made its willingness clear to receive the communication by electronic means.[94]

One might question the need for the Order when looking at section 15 of the PWA 1996. While it does not contemplate electronic service, it notes methods by which service 'may' be effected. The word 'may' is important as one could argue that it implies the list of how communication is effected is not an exhaustive one and therefore service by email, for example, is permissible.

While this is a logical conclusion, before the decision in *Knight v Goulandris*,[95] the practice was that the methods of service expressly stated in the PWA 1996 were considered as the only available ones. The Order was introduced as a reflection of the fact that, at the time the PWA 1996 was enacted, electronic methods of service were not available.

Section 15(1) of the PWA 1996 sets out the methods by which notices or documents "*may be served on a person*." It does not include email service. However, section 15(1A) (introduced by the Order) adds that electronic communication can constitute service "*but only if the recipient has stated a willingness to receive the notice or document by means of electronic communication*," has not withdrawn that statement, and the notice or document was transmitted to an electronic address specified by the recipient. Subsections (1A)–(1C) were introduced as amendments with effect from 6 April 2016 by article 2(1) of the Order pursuant to the power contained in section 8 of the Electronic Communications Act 2000.[96]

In *Knight v Goulandris*,[97] the county court held that email did not constitute valid service as it considered the methods of service listed in section 15(1) as exhaustive, justifying the need for section 15(1A). However, the Court of Appeal overturned the county court's decision, holding that the word 'may' in section 15(1) need not be read restrictively but should be read permissively as to the means by which a notice or other document could be validly served.

94 Pyramus and Thisbe Club, *The Party Wall Act Explained – A Commentary on the Party Wall Etc. Act 1996*, 3rd edn (2016); As part of the changeover to the Pyramus and Thisbe Society, a working group has been set up in 2022 to review this publication, otherwise also known as the 'Green Book'.
95 *Knight v Goulandris* [2018] 1 W.L.R. 3345.
96 ibid. 9.
97 ibid.

L.J. Patten noted:[98]

> *In my view, the most attractive way of presenting the argument against section 15(1) being merely permissive is the construction of "may" which Longmore LJ adopted in the Ener-G Holdings case [2013] 1 All ER (Comm) 1162, para 45. "May" is explicable by reference to the choice which the section gives the serving party between the specified alternative methods of service and does not therefore require to be read as permitting further unspecified methods of service that would satisfy the common law requirement of receipt. But this construction of the word "may" has been rejected in relation to section 23 of the LTA and more generally by the Court of Appeal in Hastie & Jenkerson v McMahon [1990] 1 WLR 1575. It has also failed to be accepted in a contractual context in the Ener-G Holdings case. Although one can say that those were decisions in relation to different statutes and contracts, the language under consideration in those cases was sufficiently close to that of section 15 to make that, in my view, an unconvincing ground for distinguishing those cases and they do provide at the very least highly persuasive authority at Court of Appeal level for construing section 15 in the same way. In the absence of any other circumstances or internal indications in the statutory provisions themselves, there is no justification in my view for giving "may" in section 15 a different meaning. I would therefore allow this appeal.*

3.6 Conclusion

The PWA 1996 is the result (and part of) of legislative evolution in the area of party walls over the course of several centuries. It aims to regulate relationships between the building owner and the adjoining owner who have rights and obligations in relation to the relevant party wall, and to assist with dispute prevention and resolution.

There are various uncertainties emanating from the PWA 1996 in relation to the scope and location of the relevant works performed on a party wall and construction rights.

While the PWA 1996 has no mechanism in place to force the relevant parties to comply with it, this is not uncommon in acts of Parliament. Depending on the circumstances, such obligations can still be enforced as breaches of statutory duties (or otherwise). While courts may take the non-compliant party's conduct into account and try to penalise a party for not complying with the PWA 1996 notice procedure, it is not a criminal offence not to comply with it. At the same time, if the building owner fails to trigger the mechanism under the PWA 1996 (not serving an appropriate party wall notice), it exposes itself to the possibility that the adjoining owner may start court proceedings against such a building owner, which can result in increased cost, time and emotion, compared to resolving the dispute out of court.

Finally, it is important to bear the Order in mind in the context of the PWA 1996 to ensure that any electronic communications in relation to party walls are validly served.

98 ibid. 37–38.

4

Disputes Related To Party Walls

4.1 Introduction

This chapter focusses on the dispute resolution process in relation to party walls. The analysis first looks at the procedure under the PWA 1996, then at other dispute resolution methods and the reasons for these as well as their potential outcomes. While useful and ideally the basis where to start prior to starting relevant works affecting party walls, the dispute resolution mechanism under the PWA 1996 does not provide the only viable dispute resolution method and should not be seen as the only mechanism that should be followed. The point is that, as discussed later, parties to party wall disputes have several options available to them in relation to the ways in which they can navigate the dispute resolution process and how formal or informal they would like it to be.

4.2 Notices and the Dispute Resolution Procedure Under the Party Wall etc. Act 1996

Ideally, the building owner and the adjoining owner enter into a collaborative dialogue before the building owner serves an appropriate party wall notice under the PWA 1996. Where the building owner triggers the notices mechanism under the PWA 1996 by serving an appropriate party wall notice, an adjoining owner (or adjoining owners) can agree with the building owner's proposals or come to an agreement with the building owner on changes in the way in which the works are to be performed.[99] As long as the building owner meets the requirements of the PWA 1996, it can carry out the works to which the notice regime applies under the PWA 1996. Wherever the works by the building owner fall under the PWA 1996, according to section 8 of the PWA 1996, it will have the right to enter upon the adjoining owner's land for the purpose of carrying out the works. However, such access is subject to the relevant notice given by the building owner prior to entering the adjoining owner's land.

99 DCLG's Explanatory Booklet (n 26) 4.

Party Wall Disputes: The Party Wall etc. Act 1996 and Beyond – *Legal Coherence and Dispute Management*, First Edition. Dr Laura Lintott.

4.3 When a Dispute Arises Under the Party Wall etc. Act 1996

According to section 5 of the PWA 1996, a dispute has arisen if:

a) the adjoining owner does not serve a notice within the period of 14 days on the building owner (beginning with the day on which the building owner's notice was served on the adjoining owner) consenting to building owner's notice in relation to:[100]
   i) existing party walls and structures (section 2 of the PWA 1996);
   ii) excavation and construction (section 6 of the PWA 1996);
b) the building owner does not serve a notice within the period of 14 days on the adjoining owner (beginning with the day on which the adjoining owner's notice was served on the building owner) consenting to the adjoining owner's counter-notice (section 5 of the PWA 1996).
   (deemed dispute)

A dispute also arises where the adjoining owner expressly dissents in response to the building owner's relevant notice.

As for work done with regard to a new building on the line of junction (section 1 of the PWA 1996), the building owner may only build the new wall as an external wall entirely within its own land and at its own expense if the adjoining owner does not respond after 14 days of the notice having been served by the building owner.

4.4 Notice Procedure

As for notice provisions, the building owner has to first establish whether any of the works it is planning to carry out fall under the PWA 1996. If so, the building owner should give notice to all relevant adjoining owners of the works to which the PWA 1996 applies. The notice requirements are set out below.

4.4.1 Works to an Existing Wall - Sections 2 and 3 of the Party Wall etc. Act 1996

The building owner has to give the adjoining owner two months' notice before the works commence. Alternatively, the building owner can obtain the adjoining owner's prior written consent to the works under section 3(3) of the PWA 1996.

In response, the adjoining owner can:

a) Serve a counter-notice under section 3 of the PWA 1996 within one month under section 4(2) of the PWA 1996. Unless the building owner consents to the counter-notice within 14 days of its service, a dispute will arise. However, there is no provision under the PWA 1996 for a counter-notice with regard to new walls or party fence walls under section 1 of the PWA 1996, nor with regard to excavations under section 6 of the PWA 1996. The purpose of the counter-notice is to give the adjoining owner the option to have any additional works carried out by the building owner on top of the works specified in the building owner's initial notice given under section 3 of the PWA 1996.

The counter-notice has to specify the additional works that the adjoining owner requires to be incorporated into the building owner's works and should enclose plans or other particulars of the works. In the case of *Bridgland v Earlsmead Estates Ltd*,[101] it has been established that:

---

100 ibid. 8.
101 2015 WL 3953034, 18.

> *A party structure notice relates to the building owner's proposed works. Following such service, an adjoining owner may (but does not have to) serve a counter notice. But such a counter notice relates to other work to be carried out on the "party fence wall or party structure" (see the opening phrase of section 4 (1) (a) of the 1996 Act) which "may reasonably be required for the convenience of the adjoining owner" (see the closing phrase of section 4 (1) (a) of the 1996 Act). Such other work which is the subject matter of a counter notice is not the same as the proposed work which the building owner intends to carry out, and the purpose of function of a counter notice does not concern or relate to the manner in which the building owner's proposed work is to be carried out. Further, such other work can only be carried out on the "party fence wall or party structure"; it is not the same as any further work which the adjoining owner might wish to carry out, or have carried out on his own land. A counter notice therefore cannot, and does not, relate to such further work ...*

When a counter-notice has been served, the building owner has 14 days within which to give its consent and so has to comply with the counter-notice subject to section 4 of the PWA 1996. If the building owner does not consent within 14 days of the counter-notice having been served, a dispute arises to which section 10 of the PWA 1996 applies. Any costs of the additional works under the counter-notice are the liability of the adjoining owner.

b) Give a written notice of consent within 14 days under section 5 of the PWA 1996 with regard to an existing wall under section 2 of the PWA 1996. The consent should not be a general consent but specify what exactly it relates to.[102] Once the adjoining owner has given its written consent to works under section 3 of the PWA 1996 but later suffers damage, it can pursue its rights also under section 10 of the PWA 1996 since the case of *Onigbanjo v Pearson*.[103]
c) Refuse to consent in writing. Or
d) Ignore the building owner's notice, in which case a dispute will automatically arise after 14 days from the service of the building owner's notice.

4.4.2 New Wall or Party Fence Wall Under Section 1 of the Party Wall etc. Act 1996

When building a new wall or party fence wall under section 1 of the PWA 1996, the building owner has to give one month's notice of its intention to build the new wall or party fence wall on the line of the junction, i.e. the boundary.

First, the building owner needs to follow the notice procedure under section 1 of the PWA 1996. For walls astride a boundary, the adjoining owner has 14 days within which to give its written consent. If this is given, the new wall or party fence wall can be built half on each of the owners' land with the costs being divided between the owners.

If the building owner wants to build a wall on the line of junction placed wholly on its own land, it has to serve on any adjoining owner a notice at least one month before the intended work starts indicating the building owner's desire to build and describing the intended wall under sections 1(5) and (6) of the PWA 1996. The building owner can start the works once the one month has expired from the day the notice has been served on the adjoining owner. Such

102 *Seeff v Ho* [2011] EWCA Civ 186.
103 [2008] B.L.R. 507.

works will have to be done at the building owner's expense, and the building owner will be responsible for compensating the adjoining owner and Adjoining Occupier for any damage to their property caused by the works (section 1(7) of the PWA 1996).

4.4.3 Excavations Within Three or Six Metres of the Adjoining Owner's Buildings Under Section 6 of the Party Wall etc. Act 1996

Where there are excavations within three or six metres of the adjoining owner's buildings under section 6 of the PWA 1996, the building owner has to give the adjoining owner one month's notice of its intention to carry out the excavation works (to which section 6 of the PWA 1996 applies). The building owner needs to follow the notice procedure according to section 6 of the PWA 1996. The building owner has an obligation to underpin or strengthen the foundations on the adjoining owner's land if the adjoining owner requires this. A dispute will arise if the adjoining owner does not respond to the building owner's notice within 14 days.[104] Although the PWA 1996 does not provide for any enforcement procedures, if a building owner fails to serve a notice, the adjoining owner may seek to stop the building owner's work. This can be pursued through court injunction if the building owner starts such work without having first given notice, which is compliant with the PWA 1996.[105]

4.5 Party Wall Dispute Resolution Procedure Under the Party Wall etc. Act 1996

4.5.1 Introduction

The particulars in relation to service of notices are governed by section 15 of the Order.

If the building owner fails to serve the correct notice or serves the notice on the wrong recipient, the consequences can be severe. In reality, such a failure might be an honest mistake, and the most amicable way forward is for the adjoining owner to ask the building owner to stop work and follow the correct procedures. The adjoining owner may decide, however, to seek an injunction or compensation instead.

The notice and dispute resolution mechanism under the PWA 1996 is one of the mechanisms that can be effectively applied and further below in this chapter, further dispute resolution options are explored and advocated for. Diagrams 4.10 and 4.11 below set out the notice requirements and dispute resolution mechanism under the PWA 1996 for building owners and adjoining owners, respectively. For the same reason, also the matter of adjoining owner's counter-notice has not been incorporated in Diagrams 4.10 and 4.11.[106]

---

104 Reid (n 77).

105 DCLG's Explanatory Booklet (n 26) 10.

106 As noted earlier, for work done with regard to a new building on the line of junction (section 1 of the PWA 1996), the building owner may only build the new wall as an external wall entirely within its own land and at its own expense if the adjoining owner does not respond after 14 days of the notice having been served by the building owner (DCLG's Explanatory Booklet, (n 26) 24, 33, 49, 55). Diagrams 4.10 and 4.11 were inspired by the Party Wall Company, 'Party Wall Process for Building Owners', http://partywallcompany.co.uk/PDF/The_Party_Wall_Process_For_Building_Owners.pdf and 'Party Wall Process for Adjoining Owners', http://partywallcompany.co.uk/PDF/The_Party_Wall_Process_For_Adjoining_Owners.pdf.

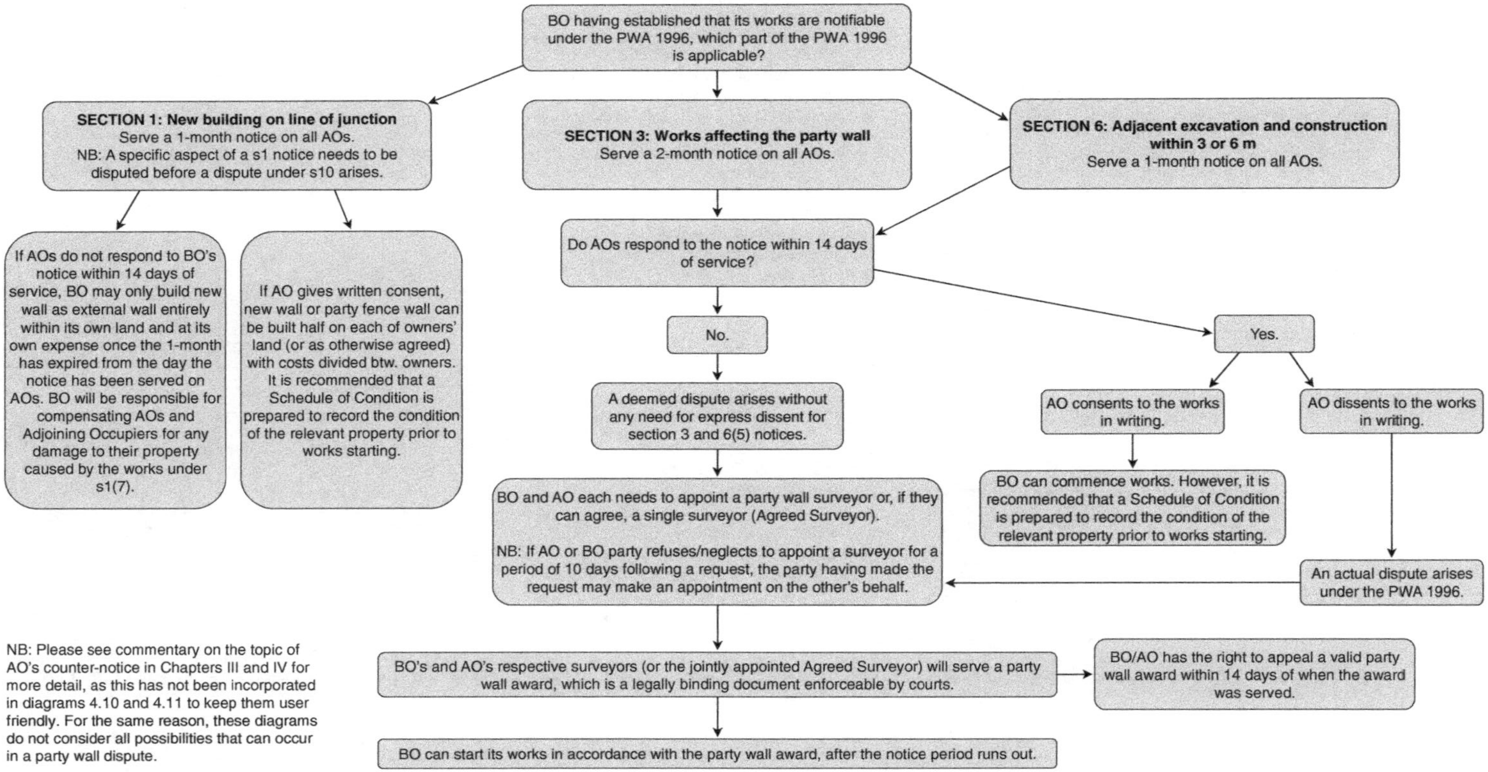

NB: Please see commentary on the topic of AO's counter-notice in Chapters III and IV for more detail, as this has not been incorporated in diagrams 4.10 and 4.11 to keep them user friendly. For the same reason, these diagrams do not consider all possibilities that can occur in a party wall dispute.

Diagram 4.10 Party wall notices mechanism for building owners.

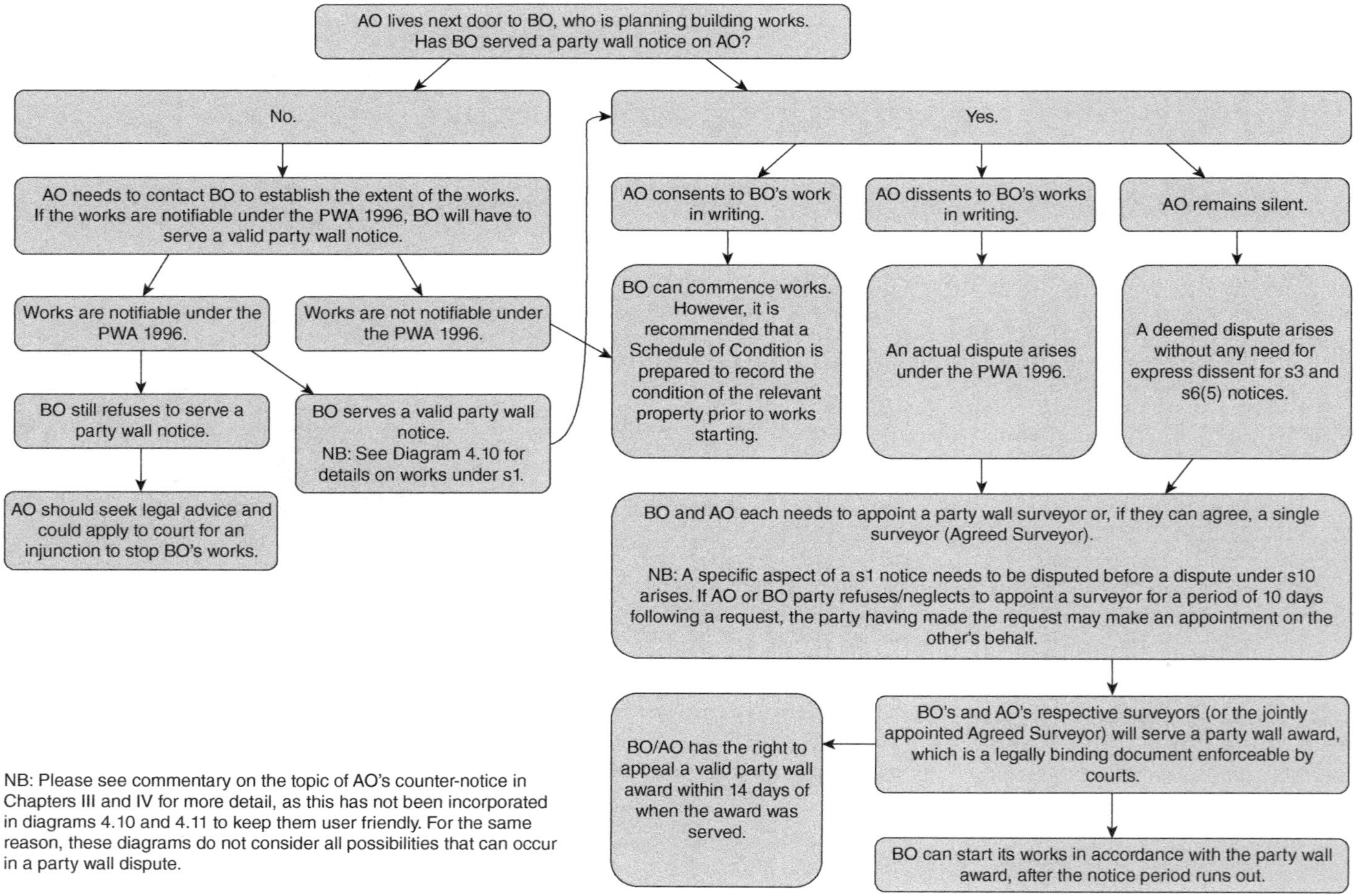

NB: Please see commentary on the topic of AO's counter-notice in Chapters III and IV for more detail, as this has not been incorporated in diagrams 4.10 and 4.11 to keep them user friendly. For the same reason, these diagrams do not consider all possibilities that can occur in a party wall dispute.

Diagram 4.11 Party wall notices mechanism for adjoining owners.

The dispute resolution procedure under the PWA 1996 is regulated by section 10 of the PWA 1996. This procedure applies where a dispute arises between a building owner and an adjoining owner regarding works to which the PWA 1996 applies.

In the case of a dispute, the points set out below apply.

a) A dispute will arise if there is no written response by the adjoining owner within 14 days (or by way of analogy, if a timely response specifies a clear dissent) to a notice served by the building owner in relation to:
   i) works to an existing party wall under section 3 of the PWA 1996; or
   ii) an excavation under section 6 of the PWA 1996.
b) A dispute will arise if the building owner does not consent to the adjoining owner's counter-notice within 14 days of service of the counter-notice under section 4 of the PWA 1996.
c) If the building owner served an appropriate party wall notice triggering the application of the PWA 1996 and a dispute arises, parties should resolve the matter as follows:[107]
   i) First level – Amicable discussion between the building owner and the adjoining owner. Any agreements should be in writing.
   ii) Second level – If an agreement cannot be reached, it is recommended that the building owner and the adjoining owner appoint an agreed surveyor to prepare an award.
   iii) Third level – If the building owner and the adjoining owner cannot agree on appointing an Agreed Surveyor, each (the building owner and the adjoining owner) can appoint a surveyor to prepare the award together. Together, the two appointed surveyors will select a third surveyor under section 10(1)(b) of the PWA 1996. The third surveyor comes into play when the other two appointed surveyors cannot agree or if the building owner or adjoining owner or either surveyor calls upon the third surveyor to make the award under section 10(1) of the PWA 1996. The surveyors' appointment should be made in writing.

The first step[108] for the party wall surveyor is to determine whether they have jurisdiction to act in order to resolve the dispute, deriving such jurisdiction in party wall matters from section 10 of the PWA 1996. This includes establishing:

a) whether the structure to which the dispute relates is a party structure or a party fence wall; and
b) whether the building owner issued an appropriate party wall notice under the PWA 1996.

The surveyor(s) does/do not have jurisdiction to make an award where the dispute in question is not a dispute for the purposes of section 10 of the PWA 1996.[109] This point

107 DCLG's Explanatory Booklet (n 26) 12.
108 Bickford-Smith, Nicholls and Smith (n 22) 8.41.
109 *Reeves v Blake* [2009] EWCA Civ 611.

was raised in the case of *Power v Shah*,[110] where it was initially held in the county court (which decision was subsequently approved both in the High Court and in the Court of Appeal) that the surveyors that were allegedly instructed under section 10 of the PWA 1996 did not in fact have jurisdiction to make an award. In this case, the building owner had done works to its property, which caused damage to the property of the adjoining owner.

The building owner never served the relevant notice under the PWA 1996. As the PWA 1996 was never invoked by the building owner, the dispute in question was not one arising under the PWA 1996:

> *... there is nothing in the Act which permits an adjoining owner unilaterally to trigger s.10 in the way that happened here. If, as Mr Isaac submitted, an adjoining owner can engage the s.10 mechanism in the absence of a s.3 notice, because of his or her unhappiness with works which have already been carried out, then I consider that such a right would have had to have been expressly set out in the Act. It would be, on any view, a significant extension of the regime that requires a prospective notice by the building owner. It is common ground that such an entitlement finds no expression anywhere in the Act.*[111]

This means that the PWA 1996 offers an alternative to court dispute resolution; however, this is subject to the building owner (the party wanting to perform the relevant works) serving the relevant notice on the adjoining owner under the PWA 1996. If the building owner does not follow the notice mechanism under the PWA 1996, neither the building owner nor the adjoining owner benefits from the PWA 1996. However, the adjoining owner has recourse to court. The Court of Appeal confirmed that the dispute resolution mechanism in section 10 of the PWA 1996 is only available where the building owner serves the relevant notice on the adjoining owner under the PWA 1996 prior to the proposed party wall works. An adjoining owner in the case had no entitlement to unilaterally invoke the PWA 1996 and the dispute resolution procedure contained within it in the absence of the appropriate party wall notice served by the building owner.[112]

4.5.2 The Surveyor

The surveyor has to act with impartiality[113] and has to make use of their professional and technical knowledge as well as knowledge of the PWA 1996. There are similarities between the surveyor's role and the role of an arbitrator, expert witness and contract administrator (under a building contract). As a quasi-arbitrator,[114] the surveyor has to abide by the rules

---

110 (n 2).
111 ibid. 31.
112 *Power v Shah* (n 2).
113 *Welter v McKeeve* (CCCL, 27 November 2018).
114 Lord Lytton discussing the Party Wall, Etc Bill in the House of Lords Debate, which led to the PWA 1996 (n 57).

of natural justice meaning that they must enable the parties to make submissions and then give them due consideration with regard to any submissions made. The jurisdiction of the surveyor is limited to:[115]

a) determining the parties' dispute about works the building owner is allowed to carry out;
b) works which fall under the PWA 1996;[116] and
c) awarding compensation and costs dealing with losses and inconvenience caused by such works.

The surveyor has no jurisdiction over works carried out by the building owner outside of the scope of the party wall award, nor to retrospectively approve work that has been completed. The surveyor can make a further award dealing with unforeseen events, if necessary, in accordance with section 10(12) of the PWA 1996.

Under section 10(5) of the PWA 1996, a party wall surveyor can be replaced only if they die, become incapable of acting or deem themselves incapable of acting (all of which also applies to the third party wall surveyor under section 10(9) of the PWA 1996). The decision in *Mills v Savage*[117] confirmed that a surveyor is entitled to declare themselves incapable of acting for whatever reason they think fit, irrespective of actual incapacity. It was also held that it was no business of the other appointed/selected surveyors to police that declaration.[118] The appointing party or the other two surveyors may in such circumstances appoint another surveyor. Once the third surveyor has been selected, it is not open to party-appointed surveyors to replace the third party wall surveyor, except where the third party wall surveyor refuses to act, neglects to act, dies or deems themselves incapable of acting.[119]

If a party wall surveyor refuses under section 10(6) of the PWA 1996 or neglects under section 10(7) of the PWA 1996 to act effectively, the other may proceed in that surveyor's absence (acting *ex parte*). This also applies to the third surveyor (section 10(9) of the PWA 1996). According to section 10(7), a request has to be served on the surveyor who neglects to act effectively giving them ten days in which to act. If there is more than one surveyor, as long as the surveyor responds before the other surveyor acts *ex parte* (even if it is after such ten days), this is deemed to be acceptable. As for section 10(6) of the PWA 1996, there is no need to serve such a request. Note that if a surveyor wrongly acts *ex parte*, a court will quash such an award.[120]

A party wall surveyor may rely on advice from others, especially in matters requiring expertise, but the surveyor should disclose to the parties when relying on such advice and from whom. Engineering advice and quantum surveyor costings are likely to form the most common examples of taking advice in practice, but a range of specialist advice

115 Reid (n 77).
116 *Woodhouse v Consolidated Property Corp Ltd* [1992] 11 WLUK 28.
117 (CCCL, 15 June 2016).
118 N Isaac, https://www.boundariesbook.co.uk/?p=105.
119 *Reeves v Young Young & Antino* (CCCL, 3 January 2017).
120 *Patel v Peters* [2014] EWCA Civ 335.

may be required, including legal advice, particularly when the award is covering matters of compensation. At the same time however, the party wall surveyor should not take advice when it is not needed as this will increase the cost to the paying owner unnecessarily.[121]

Party wall surveyors and lawyers should ensure they have knowledge of the PWA 1996, case law, legislation and other matters (such as boundaries, easements and tort) referred to in this book (and beyond, keeping up to date with latest developments), health and safety legislation, building regulations and civil procedures etc.

Owners rely on the advice given by their respective party wall surveyor as to construction issues and lawyer as to the legal issues. It is important that the surveyor explains to the appointing owners the limitations of the PWA 1996 and where the surveyor's jurisdiction starts and ends. Separate legal advice should be sought by the appointing owners on matters outside the surveyor's jurisdiction to ensure the owners are aware of all their rights and obligations. This does not prevent the relevant party from also seeking legal advice on the PWA 1996 itself as the relevant party will benefit from the different perspectives that lawyers and party wall surveyors can provide. It is also of great importance that party wall surveyors and lawyers knowledgeable in the area of party walls and related disputes operate on a collaborative basis. Lawyers and party wall surveyors each have a distinct important role and expertise which complement each other. Those affected by issues linked to party walls should seek legal advice early (ideally before a dispute arises) to be fully aware of their rights and obligations and timing of the need to appoint a party wall surveyor. This will help minimise cost, time and emotion wastage should a party wall dispute ensue – or to even prevent such a dispute.

HHJ Edward Bailey summarised five qualities that he suggests apply to the party wall surveyor acting in their dispute resolution capacity:[122]

... namely the duty to act:

- *(i) intra vires, that is within the scope of the surveyor's powers;*
- *(ii) impartially, with a lack of bias (not inconsistent with ensuring that the appointing owner's interest are protected);*
- *(iii) so as to ensure that each owner, usually through that owner's appointed surveyor, has the opportunity to put forward for consideration any matter relevant to the making of the award;*
- *(iv) so as to ensure that both owners' cases are given proper consideration;*
- *(v) thoroughly, so as to ensure that all necessary investigations are carried out before making an award and that the award is made having regard to all the material obtained.*

If the PWA 1996 applies and a dispute arises, the surveyor plays an important role in a dispute resolution mechanism outside of the court room, saving cost and time.

---

121 HHJ Edward Bailey, 'Proper conduct for the party wall surveyor', A Faculty of Party Wall Surveyors Webinar – 12 November 2020, 58–59.

122 ibid. 7.

4.5.3 The Award

According to section 10(12) of the PWA 1996, the surveyor's award may determine:

a) the building owner's right to execute any work;
b) the time and manner in which the building owner executes any work; and
c) any other matter arising out of or incidental to the dispute in question (including the costs of making the award).

According to paragraph 51 of DCLG's Explanatory Booklet, the award is also to include a provision in relation to any necessary checking that the work has been carried out according to the award. Although there is no provision in the PWA 1996 as to a prescribed form for the award, in practice many party wall surveyors use the model party wall award prepared by the Royal Institute of Chartered Surveyors (**RICS**).[123]

According to paragraph 14 of DCLG's Explanatory Booklet, the award should state:

a) what works will be carried out;
b) when and how the works will be carried out, with timing requirements in the award overriding the notice provisions in the PWA 1996;
c) any additional works required;
d) any other matter such as costs of making the award; and
e) a record of the condition of the adjoining owner's property before the works begin (often with photographic evidence) to protect the adjoining owner (**Schedule of Condition**).

Party wall surveyors may not delegate the making of an award.[124] From a practical perspective, before any work starts, a record of condition should be included (universally referred to by party wall surveyors as a 'Schedule of Condition'), and this should cover all sections of the adjoining owner's property on which the works could impact. Without a Schedule of Condition, it would be very difficult for the adjoining owner to prove any damage caused by the building owner's work.[125] A Schedule of Condition is helpful in avoiding arguments later as to whether damage was pre-works or consequent on the works.

Deemed service in relation to a party wall award is not ascertained by the application of the Civil Procedure Rules (**CPR**).[126] In *Knight v Goulandris*,[127] it was stated that:

> *... section 15(1)(a) plainly requires such service on a person, for it speaks of the method by delivering it to him in person. I find it hard to conceive that such a method does not*

123 The model party wall award template is appended to: RICS, 'Party wall legislation and procedure – England and Wales,' (7th edn, 2019).
124 *Longmire v Maldura* (CLCC, 29 January 2015).
125 Practical Law Construction,'PWA 1996: reviewing the party wall surveyor's award' (Maintained) *Practical Law UK Practice Note*, Resource ID 7-516-5148, https://uk.practicallaw.thomsonreuters.com/7-516-5148?originationContext=document&transitionType=DocumentItem&contextData=%28sc.Default%29&comp=pluk.
126 The White Book Service 2024 (Sweet & Maxwell, 2024).
127 (n 95), 20.

> *involve receipt by that person. Of course, such a person may decline receipt by casting it from him, but if a notice is delivered by person to another person, I do not see that it can be properly said that the person to whom the notice is delivered can say that he has not received it ...*

The principle stated in the case shows that whether a document has been served or not depends on the question of receipt of that document. The document has been served on the date of receipt occurring. While there are exceptions to this rule, the basis where to start is the point in time of the date of actual receipt and then calculating 14 days from such date for a potential appeal under section 10(17) of the PWA 1996.

A party wall award should not be confused with an arbitration award as the procedure differs in the sense that, to be able to have an arbitration award, there must be a binding arbitration 'agreement in writing' (as defined in section 5 of the Arbitration Act 1996; Arbitration Act 2025 having received Royal Assent in February 2025) between the parties (except in cases of statutory arbitration, see chapter V – HS2 Legislation). The Arbitration Act 1996 does not apply to party wall awards falling under the PWA 1996. If either of the parties is dissatisfied with the award, they have the option of appealing to the county court within 14 days of the service of the award under section 10(17) of the PWA 1996.

The decision in the case of *Zissis v Lukomski*[128] clarified what the procedure should be regarding an appeal to the county court under section 10(17) of the PWA 1996. An appeal is governed by CPR Part 52. Overwhelmingly, the appeal will be a rehearing (using the express power in the county court provided by section 10(17)), but occasionally it will make sense for the appeal to be a review. There are some situations where there cannot be an appeal, for example, where the 14-day limit for an appeal has expired. Alternatively, because the award should never have been made (for example, where the surveyor acted *ultra vires* or an error in the statutory procedure occurred). It will be a nullity, and one cannot realistically appeal a nullity. In such situations the owner complaining about the award must start CPR Part 7 (rarely Part 8) proceedings. In this case, the appellant (Zissis) appealed to the county court against an award under section 10(17) of the PWA 1996 (a statutory appeal governed by Part 52 of the CPR).[129]

Sir Peter Gibson noted that:

> *On the contrary I think it plain that Part 52 was intended to cover a form of statutory appeal like that under section 10(17) and that the provisions of Part 52 are amply sufficient to allow justice to be done on such an appeal ....*[130]

In order to appeal, the appellant needs to file a notice within 14 days of service of the award.

In the absence of an appeal, awards can only be challenged on the basis of vires/ jurisdiction, whether in proceedings for a declaration as to the absence of such vires/ jurisdiction, or in response to proceedings brought attempting to enforce an award.[131]

---

128 [2006] EWCA Civ 341, 40.
129 ibid.
130 ibid., 39.
131 *Arbitration between Stone and Hastie, Re* [1903] 2 K.B. 463; *Gyle-Thompson v Wall Street Properties Ltd* [1974] 1 W.L.R. 123; *Woodhouse v Consolidated Property Corp Ltd* (n 116).

A separate action can go ahead by way of a claim under CPR Part 7, or by way of the alternative procedure under Part 8. A Part 7 claim is more usual as most challenges of the validity of an award depend on showing evidence that part or all of the award is *ultra vires*. This would include a situation where the challenge is that the supposedly authorised works go beyond the powers of the surveyor under section 2 of the PWA 1996. In addition, it will be necessary to show to the court what the award purports to authorise by reference to the documents referred to in the award. Similarly, in a situation where it is to be determined whether the costs have been awarded properly, factual evidence will have to be drawn upon to show the circumstances in which the costs were incurred. On the other hand, a claim under Part 8 focusses on questions of law without pleadings and with limited evidence. This would work where, for example, an award is challenged on the basis of a clear procedural irregularity. However, where there are fact-specific challenges and cases, where pleadings and cross-examination of witnesses are necessary, a claim under Part 8 is not suitable and a claim under Part 7 should be sought.[132]

The PWA 1996 is silent on the point as to whether, whilst the parties await the result of the appeal, the award is binding on the parties or whether the award is suspended. The general rule (for judgments and applied by analogy to awards) is that the award is binding whilst awaiting the decision on the appeal. The PWA 1996 would fail to operate correctly if it were allowed for the adjoining owner to delay the implementation of the award by filing a notice of appeal. This does not, however, leave, for example, the adjoining owner helpless in a situation where it wishes to prevent the consequence of an award. In such a case, the adjoining owner has the option to apply for an interim injunction and give a cross-undertaking in damages.[133] The legal test for the granting of an interim injunction[134] includes having a good arguable case, that damages would be an inadequate remedy and that the balance of convenience favours the grant of an injunction.

In a situation where a part of an award is invalid, this does not invalidate the award in its entirety but only where the invalid part can be severed and is not intertwined with the rest of the award[135] so that severing the invalid section of the award would not distort the award.[136] The court will therefore consider the severability of the part of the award which is invalid.[137]

Under section 19 of the Senior Courts Act 1981, the High Court has jurisdiction to grant a declaration as to the parties' rights (a discretionary remedy).[138] This jurisdiction is not ousted by the dispute resolution mechanism under the PWA 1996.[139] The court will,

---

132 Bickford-Smith, Nicholls and Smith (n 22) 10.8.

133 ibid. 10.15. In addition, often as a 'shortcut', an appellant in a party wall appeal will ask the court for a 'stay' of the award in the appellant's notice. Not infrequently, the court will grant such a stay without requiring a cross-undertaking to be given.

134 *American Cyanamid Co v Ethicon Ltd* [1975] A.C. 396.

135 *Selby v Whitbread & Co* (n 6), applied in *Gray v Elite Town Management Ltd* [2016] 11 WLUK 91.

136 *James E McCabe Ltd v Scottish Courage Ltd* [2006] EWHC 538 (Comm).

137 Bickford-Smith, Nicholls and Smith (n 22) 10.17.

138 *Lea Valley Developments Ltd v Derbyshire* [2017] EWHC 1243 (TCC), 30: "*The court has an inherent jurisdiction to provide declaratory relief under s.19 of the Senior Courts Act 1981 and as set out in CPR 40.20*"

139 Bickford-Smith, Nicholls and Smith (n 22) 15.41.

however, not always exercise this power and will take into account individual circumstances of the matter before it in deciding whether it will exercise its power under its jurisdiction. Courts will not support a claim for declaratory regime if they view it as the party's way to subvert the statutory regime (according to practitioners' views).[140]

In *Lea Valley Developments Ltd v Derbyshire*,[141] O'Farrell J. explained what the court's reasons were when ruling that it should exercise inherent jurisdiction:[142]

> *I then turn to the issue as to whether or not the court should exercise its inherent jurisdiction in this case or whether it would be more appropriate to stay the proceedings pending determination by the surveyors of the outstanding matters in relation to the party wall dispute. ... I start by looking at the award that was issued on 19 December 2014. This award is in fairly clear and straightforward terms. ... The option of making payment in lieu of carrying out the works is one that can be exercised at the discretion of the defendant and is not something that can be imposed on the defendant. That award is final and conclusive in accordance with s.10(16) no appeal having been started by either party to overturn or modify the terms of that award. The issue that the claimant is asking the court to determine is one that is narrow in compass and one that ought to be capable of determination by the court after a relatively short hearing. ... it would not be trespassing on the role of the surveyors under the Act, it would simply be identifying the appropriate test that the surveyors should adopt when making their assessment. The court would not be determining the dispute as to the amount of compensation, it would simply be determining the appropriate interpretation of the law, the Award and the Act so as to enable the surveyors to move forward and settle the further award. ... The parties have not yet referred this matter to the surveyors, and therefore the sensible course is for both parties to wait for the court determination before seeking a further award from the surveyors. The advantage of having this point determined is that it could narrow the issues between the parties and leave the surveyors to determine the facts and figures based on the reports that are already available. Although that will not exclude the possibility of any appeal, it is very likely to dispose of the key dispute. Given that this matter can be determined by the court in short order, it is the sort of issue that is appropriate to be dealt with by way of a Part 8 claim.*

It should be noted that the process of claiming declaratory relief is not a straightforward one.

Even if the party wall surveyors agree on a method statement, this does not mean that the adjoining owner will not sustain any damage to its property. The agreed method statement will also not protect the building owner from the adjoining owner's potential claim for compensation (section 7 of the PWA 1996). The PWA 1996 also covers the costs linked to

---

140 Reid (n 77). In *Gray v Elite Town Management Ltd* (135), the county court refused declaratory relief. In this instance, this was not on the basis that doing so would be subverting the statutory regime, but rather that the declaration would have no point beyond academic interest.
141 (n 138).
142 ibid. 32.

making the party wall award. Costs are covered in section 10 of the PWA 1996, and expenses are covered in section 11 of the PWA 1996. Neither costs nor expenses are defined terms in the PWA 1996.

4.5.4 Costs

As costs are noted a number of times under section 10 of the PWA 1996, the term is clearly used in a similar way as it is also used in relation to litigation. Therefore, costs under the PWA 1996 include:

a) fees of any surveyor appointed to discharge a function under the PWA 1996; and
b) legal and other professional costs incurred by the parties as to operating the procedures of the PWA 1996.[143]

Costs could include, for example, those incurred in relation to service of notices, fees of engineering and other consultants and fees for legal advice related to the procedures under the PWA 1996.[144]

Where the two appointed surveyors disagree, the third surveyor may hold a hearing at which to address the relevant issue and may make an award. The costs of such a hearing would be regarded as costs payable under the PWA 1996.[145]

According to section 10(12)(c) of the PWA 1996, the party wall surveyor has jurisdiction to decide who will pay for the costs of making the award. Section 10(13) of the PWA 1996 relates to the terminology of 'reasonable costs' and includes the costs in:

a) making and obtaining the award under section 10(13)(a) of the PWA 1996;
b) inspecting the works to which the award relates under section 10(13)(b); and
c) dealing with any other matter arising out of the dispute under section 10(13)(c) of the PWA 1996.

Therefore, the fees of the party wall surveyor fall within section 10 of the PWA 1996. In *Farrs Lane Developments Ltd v Bristol Magistrates' Court* it was held that:[146]

> *The statute enables the surveyor(s) to determine which party is to pay the costs awarded without limiting the discharge of that obligation in the award to a payment to another party, rather than to the surveyor entitled to receive that payment.*

What this is means is that the party wall surveyor can sue/enforce the payment directly.

If the building owner objects to the costs claimed by an adjoining owner, the adjoining owner can refer the same to the third surveyor (or the party-appointed surveyors can make an award as to reasonable costs). The PWA 1996 imposes a ceiling of reasonableness on those costs.

---

143 Bickford-Smith, Nicholls and Smith (n 22) 11.3.
144 *Onigbanjo v Pearson* (n 103).
145 *Chartered Society of Physiotherapy v Simmonds Church Smiles* [1995] 1 E.G.L.R 155.
146 [2016] EWHC 982 (Admin), 49.

The costs of the award can include legal costs as long as these are reasonably and properly incurred, and are in relation to the dispute. However, costs of actual or threatened proceedings for the purposes of enforcing common law or equitable remedies (for example, damages, injunction to restrain works or for trespass or nuisance)[147] fall outside of the PWA 1996.[148] Although such legal costs fall outside of the PWA 1996, if court proceedings are subsequently issued, those legal costs may be recoverable under the usual CPR costs principles. The PWA 1996 does not cover interest payable for late payment of the party wall surveyor's costs; however, awards may do so.

According to section 10(15) of the PWA 1996, the third party wall surveyor can require that their costs are paid before serving the award on either the parties or their party wall surveyors. There is however a limit in terms of when the third party wall surveyor can require their costs, as they cannot do so before they even start work or before beginning to make an award.[149]

4.5.5 Expenses

Expenses are referred to in sections 1(3)(b), 1(4)(a), 1(7), 11, 12, 13 and 14 of the PWA 1996. The term 'expenses' refers to the actual cost of carrying out works including the prime cost of the works, reasonable incidental professional fees (for example, architect or quantity surveyor), fees paid to statutory authorities and insurance related to the works.[150]

In cases where the works to a party wall are part of a wider project, it is necessary to ensure that the expenses are separated in relation to the works to which the PWA 1996 applies from the total cost of the project.[151] Costs in relation to issues which are too remote from the relevant works falling under the PWA 1996 will not form any part of the expenses under the PWA 1996 (for example, feasibility studies, environmental assessments or expenses in relation to the obtaining of a planning permission).[152]

If the adjoining owner is responsible under section 11 of the PWA 1996 to pay for some of the costs and expenses of the works (either as this was agreed by the parties, or as this was decided in a party wall award), then the building owner has to serve a detailed account. This is to be done within 2 months after completion of the works on the adjoining owner showing the amount payable by each of the parties. The adjoining owner then can serve a notice of its objections within one month; otherwise, it will be deemed to have accepted the account (under section 13 of the PWA 1996). All property rights in the works remain vested in the building owner until the adjoining owner settles the account under section 14 of the PWA 1996.

Under section 17 of the PWA 1996, any sum payable under the PWA 1996 can be recovered summarily as a civil debt. This provision allows proceedings to be brought in the magistrates' court under the Magistrates' Court Act 1980.

---

147 *Reeves v Blake* (n 109).
148 Bickford-Smith, Nicholls and Smith (n 22) 11.3.
149 *Mills v Savage* (n 117).
150 Bickford-Smith, Nicholls and Smith (n 22) 11.9–11.10.
151 ibid. 11.7.
152 ibid. 11.8.

4.5.6 Compensation

Under section 7(1) of the PWA 1996, the building owner has to ensure that it does not cause unnecessary inconvenience to any adjoining owner or any Adjoining Occupicr when exercising its right. Under section 7(2) of the PWA 1996, the building owner has to compensate any adjoining owner as well as any Adjoining Occupier for any loss or damage resulting from work performed under the PWA 1996. These provisions are in place to provide a balance considering the PWA 1996 also protects the building owner. The PWA 1996 is allowing the building owner to act in a manner that would usually amount to trespass or nuisance, had it not been for the PWA 1996. This needs to be seen in context of the potential 'right/obligation to repair' as further discussed in Chapter VI of this book. Adjoining Occupiers are covered by section 7 of the PWA 1996 as well. Having noted this, Adjoining Occupiers cannot participate in section 10 of the PWA 1996 procedure but, as they may be affected by the building owner's works, section 7 of the PWA 1996 adds a safeguard to the Adjoining Occupiers, which may include tenants as well as other residents of the adjoining owner's property. However, if the building owner's works cause a loss to the adjoining owner or Adjoining Occupier and such works go beyond the works authorised by the party wall award or the PWA 1996, then the adjoining owner's or Adjoining Occupier's remedy may be a common law claim for damages in trespass or nuisance in addition to or instead of a claim under the PWA 1996 (depending on the circumstances of each individual case).

The right to compensation is not limited to compensating the adjoining owner or Adjoining Occupier for unnecessary inconvenience under section 7(1) of the PWA 1996. An adjoining owner or Adjoining Occupier is entitled to recover compensation for: a) all loss or damage resulting from the works (irrespective of whether out of unnecessary inconvenience or otherwise); and b) any loss or damage arising out of the very existence of the finished works carried out in pursuance of the building owner's rights under the PWA 1996, and not works carried out in breach of an award or outside of the scope of the building owner's rights.[153] However, if the building owner fails to serve the appropriate notice, this does not mean it succeeded in avoiding responsibility for unnecessary inconvenience caused to an adjoining owner as a result of works that would have been covered by the PWA 1996, had the building owner served the requisite notice.

In *Bridgland v Earlsmead Estates Ltd*,[154] HHJ David Grant stated

> *I therefore reject Mr Taylor's submission that the consequence of a failure to serve a party structure notice is that a building owner is not then "exercis(ing) any right conferred on him by this Act," and as a result is absolved from the requirement not to cause unnecessary inconvenience to an adjoining owner ...*

According to section 17 of the PWA 1996, all compensation payments under the PWA 1996, expenses and costs are summarily recoverable as civil debts. Section 7(1) of the PWA 1996 contains the building owner's obligation not to cause unnecessary inconvenience to any adjoining owner or Adjoining Occupier. Section 7(2) covers all compensation for 'any loss or damage' which includes loss or damage suffered as a consequence of unnecessary inconvenience under section 7(2).

---

153 Reid (n 77).

154 (n 101) 23.

In the case of *Lea Valley Developments Limited v Derbyshire*,[155] Adrian Williamson QC (sitting as a Deputy High Court Judge) analysed a situation where a party wall award was made prior to works commencing, noting that the building owner:[156]

> *Make good all structural or decorative damage to the Adjoining Owner's property occasioned by the works ... If so required by the Adjoining Owner, make payment in lieu of carrying out the works to make the damage good, such sums to be determined by the Agreed Surveyor.*

It was held, obiter, that there is a difference as to how section 2 of the PWA 1996 is treated in this context compared with the treatment of section 6 of the PWA 1996. It was noted that section 2 of the PWA 1996 made reference to 'making good', whereas section 6 of the PWA 1996 did not.[157] An obligation to make good only applies to certain notifiable works (as detailed in section 2(3) of the PWA 1996 and following subsections). The corresponding right on the part of the adjoining owner to elect for money in lieu of making good (under section 11(8) of the PWA 1996) can only apply where there is a primary obligation to make good. No such primary obligation applies to section 6 of the PWA 1996 (adjacent excavation and construction).

4.5.7 Jurisdiction to Award Compensation

The party wall surveyor's jurisdiction to award compensation under the PWA 1996 is primarily under section 7(2), and also under section 11(8). Compensation is to be assessed in the same way as damages in common law tortious claims.[158] The party wall award sets out the works that the building owner can carry out without unnecessarily affecting the adjoining owners and Adjoining Occupiers negatively. Even if the building owner follows the award correctly, loss or damage can sometimes still occur to the adjoining owner or Adjoining Occupier. In such a situation, the party wall surveyor can provide for an addition to their initial award with a further award requiring the building owner to pay compensation.[159] Where the building owner fails to comply with a party wall award during carrying out of its works resulting in loss/damage suffered by the adjoining owner, the adjoining owner can claim for remedy if it decides to start court proceedings.

4.5.8 Security

Section 12 of the PWA 1996 gives the building owner and the adjoining owner the right to ask the other party for security in relation to expenses if certain circumstances apply.[160]

155 [2017] EWHC 1353 (TCC).
156 ibid. 7.
157 ibid. 21–23.
158 ibid.
159 *Onigbanjo v Pearson* (n 103).
160 Bickford-Smith, Nicholls and Smith (n 22) 12.1.

The statutory rights given to the building owner may authorise him to embark on extensive and radical works intimately affecting the adjoining property, but his entitlement is not conditional on financial substance, or even solvency. Without any provision for some kind of monetary security the adjoining owner might find himself having to fund from his own pocket expensive repairs to his own property, caused by insolvency of the building owner supervening in the middle of his operations. The adjoining owner's right is to call for security to be provided before the works are begun, and if it is awarded, the building owner will have to provide it before beginning. Section 12(2) gives reciprocal rights to the building owner. The need for this is less obvious, but since the adjoining owner can impose additional works upon the building owner in his counter-notice (s4(1)), and there are several other ways in which he may incur liability under the Act (Chapter 11), such liabilities can be substantial enough to justify requiring security from him. What is somewhat strange is that the building owner is also given a right (which is not new) to claim security merely because the adjoining owner has claimed it from him (tit-for-tat).[161]

According to section 12(1) of the PWA 1996, the adjoining owner may serve a notice asking the building owner for security before the building owner starts the works. In *Kaye v Lawrence*,[162] Ramsey J, sitting as a county court judge, noted:

It is evident from the provisions relied on by Mr Hutchings as giving rights, such as section 2(2)(f) that, even within those provisions, all the relevant work could be carried out on the property of the building owner. In such circumstances I do not consider that it could be argued that security would only apply to work being carried out on the adjoining owner's land. For instance, cutting into the party structure on the building owner's land could weaken the structure and cause damage to the structure on the adjoining owner's land ...

Therefore, a request for security for work to which the PWA 1996 applies does not have to be only in relation to work carried out on the adjoining owner's land. On the contrary, security can be required in relation to work carried out on the adjoining owner's land as well as on the building owner's land, as long as the work falls under the PWA 1996.

4.5.9 Failure to Comply with the Party Wall etc. Act 1996

The building owner has statutory obligations under the PWA 1996 and if it does not serve the relevant notice under the PWA 1996, both the building owner and the adjoining owner will lose the PWA 1996's protection. According to the decision in *Louis v Sadiq*:[163]

The scheme under the London Building Acts (Amendment) Act 1939 was clear: the building owner has certain express rights in respect of party structures under

161 ibid. 12.2–12.3.
162 (n 61) 54.
163 (1997) 74 P. & C. R. 325, 326.

> *section 46, but these can only be exercised (1) with the adjoining owner's written consent or (2) in accordance with a valid award by the surveyor(s) appointed under section 55. The authorities referred to established that the appellant would not have been liable in nuisance if he had given notice or obtained consent in accordance with the Act and had done nothing more than was agreed or approved by the surveyors. The issue raised was whether the appellant's liability at common law was either excluded or reduced by the provisions of the Act, which he invoked after the nuisance had arisen. The adjoining owner's common law rights are supplanted when the statute is invoked, but if the building owner commits an actionable nuisance without giving notice and without obtaining consent, he cannot rely upon a statutory defence under procedures with which ex hypothesi he has failed to comply.*

In general, courts tend to side with the adjoining owner where it suffers damage caused by works where no notice was served[164] as it is only the building owner who has the power to trigger the applicability of the PWA 1996.

In *Crowley (t/a Crowley Civil Engineers) v Rushmoor BC*,[165] HH Judge Thornton QC held that:

> *... any failure to serve the requisite notice before work started would amount to a breach of statutory duty which would allow a court to award damages representing the compensation that would have been awarded by the surveyors appointed under the Party Wall Act for any damage caused by the work that would have been avoided had the notice provisions of the Party Wall Act been complied with.*

If the building owner fails to carry out the relevant works in compliance with the PWA 1996, it may be liable also for breach of statutory duty and the adjoining owner is then entitled to claim damages. One could argue that, since the claimant has a claim in tort for trespass and/or nuisance, it is unnecessary to decide whether there is also a claim for breach of statutory duty.[166]

A building owner who fails to comply with the PWA 1996 will only be left with its common law rights. The building owner will be in breach of the PWA 1996 if it fails to serve a notice, fails to serve the correct notice or serves the notice on the wrong recipient. At first instance, as it may be a matter of genuine mistake, parties should try and resolve this amicably amongst themselves. However, if needed, the adjoining owner may seek an injunction or damages.

While courts are able to award damages, as noted by HHJ Thornton QC in the above quotation, there are several aspects to this. First, the adjoining owner needs to bring a claim in court. Secondly, the court can decide, at its discretion, whether it considers it appropriate to award damages, injunctive relief or specific performance (depending on the facts of each case) – if any at all. Thirdly, while courts will generally sympathise with the adjoining

---

164 *Roadrunner Properties Ltd v Dean* [2003] EWCA Civ 1816.
165 [2009] EWHC 2237, 104.
166 *Roadrunner Properties Ltd v Dean* (n 164).

owner where it is a victim of damage caused by works performed by the building owner, which failed to serve the appropriate notice under the PWA 1996,[167] there is no guarantee that the relevant court will actually grant the relief the adjoining owner is claiming. Finally, if the adjoining owner cannot proceed with such a claim (for example due to time/financial constraints), the building owner will not suffer consequences of its legal wrong. However, if the building owner fails to serve an appropriate party wall notice triggering the mechanism under the PWA 1998, should a dispute arise, it exposes itself to the risk of the adjoining owner going to court, which is likely to be more time, cost and emotion consuming than the dispute resolution process offered under the PWA 1998.

4.5.10 Injunctions

Injunctions function as orders for a particular person or persons to perform or not to perform certain acts. Should a party be in breach of an injunction, the remedy for such a breach includes committal for contempt, a fine or sequestration of assets.[168]

Examples of injunctions with regard to the PWA 1996 include those set out below.

a) *Udal v Dutton*,[169] where HHJ Coulson QC (at the time) granted an injunction preventing the building owner from continuing to demolish the party wall in question. The building owner also had to erect a temporary hoarding where the initial party wall had already been demolished. It was held:[170]

> *The relevant principles are to be found in the older case of American Cyanamid Co.-v-Ethicon Ltd. [1975] AC 396 (HL) and in the more recent decision of the Court of Appeal in Zockoll Group Ltd.-v-Mercury Communications Ltd. (No. 1) [1998] FSR 354 and in the decision of Chadwick J (as he then was) in Nottingham Building Society-v-Eurodynamics Systems Plc [1993] FSR 468 at 474. ... The Claimants need to demonstrate three things: that there is a serious issue to be tried; that the balance of convenience favours the grant of an injunction; and that damages would not be an adequate remedy.*

b) *Histed v Prosperity Developments Ltd*,[171] where the adjoining owner was successful in its application for a without notice injunction to prevent the building owner from carrying out piling, excavation and ground works over the August bank holiday weekend. Although the parties agreed to have the injunction in place until the planning permission was acquired for the revised works, the building owner applied to vary the terms of the injunction or to have it lifted. The adjoining owner opposed the building owner's application noting that the building owner heavily disregarded its obligations under the PWA 1996 as well as

167 ibid.
168 Bickford-Smith, Nicholls and Smith (n 22) 3.51.
169 [2007] EWHC 2862 (TCC).
170 ibid. 13 and 14.
171 [2013] EWHC 4463 (QB).

the original planning permission. It was held that the injunction should remain in place. Mr Justice Holroyde held:[172]

> *I am afraid I am led to the conclusion that there is a clear inference to be drawn here of a cynical plan to try to get the job done over the Bank Holiday weekend, in the hope that it would be impracticable at such a time for neighbouring property owners effectively to prevent the work, and in the further hope that once the job had been done it would be too late for anyone to complain about this. This, as I say, is not something which happened by inadvertence or through over- enthusiasm or through ignorance of the Defendants' obligations. It was deliberately done, in my judgment, despite clear notice of their obligations.*

c) *Heathcote v Doal*,[173] where the HH Judge Grant granted an interim injunction without notice. Conduct of the parties was also an important factor in making the decision where HH Judge Grant held:[174]

> *... the whole point of the Party Wall Act is to provide for a regime of the service of notices and counter-notices, the appointment of surveyors, the provision of an award, all of which is intended to be done before works are commenced. The failure by the defendants to comply with the provisions of the Act was a matter which in my judgment did take this case and its circumstances out of the norm. I regard this as a serious matter.*

and

> *. . . The norm is to comply with the Act and serve the relevant appropriate notices.*[175]

HH Judge Grant awarded the claimants their costs of the injunction application, including the costs on an indemnity basis[176] for part of the relevant period (which were ordered because Doal failed to comply with the PWA 1996).[177] The court also criticised the first surveyor's notices, which were served after work had started and contained inaccurate and inconsistent descriptions of the proposed works, which was considered wrong and misleading.[178]

d) In *Chliaifchtein v Wainbridge Estates Belgravia Ltd*,[179] HHJ Coulson QC (at the time) awarded an adjoining owner the costs of obtaining an injunction preventing a building owner from breaching a party wall award under the PWA 1996 (by preventing excavations). The case shows what factors the court takes into account when deciding costs. The court looked at the adjoining owner's as well as the building owner's conduct.

---

172 ibid. 23.
173 (n 93).
174 ibid. 40.
175 ibid. 44.
176 ibid. 46.
177 ibid. 40.
178 ibid. 41.
179 [2015] EWHC 47 (TCC).

Unintended financial consequences of an injunction were highlighted in *Nelson's Yard Management Co v Eziefula*,[180] where the building owner failed to reply to four letters of the adjoining owner and the adjoining owner's surveyor was not able to gain access to the site. Most importantly, there was a real danger that the adjoining owner's property was at risk.[181]

Compensation may be more appropriate (rather than an injunction) where, for example, the building owner's works have already reached completion, in which case the adjoining owner may seek compensation where the building owner has failed to serve a notice. In *Crowley (t/a Crowley Civil Engineers) v Rushmoor BC*,[182] HHJ Thornton QC identified three ways in which an adjoining owner can seek compensation:[183]

> *103. Firstly, the relevant arbitration provisions provided for by the Party Wall Act can always be operated retrospectively. These provisions involve the appointment of surveyors to resolve disputes arising in connection with any matter connected with any work to which the Party Wall Act relates. The surveyors so appointed would have jurisdiction to award appropriate compensation for any damage resulting from excavation or demolition work close to the flank wall and the adjoining planter which could and should have been, but had not been, made subject to an appropriate award prior to work starting and which undermined and damaged the foundations and the property that they supported (see sections 7(2),10(1), 10(12), 10(13)(c)) and 17 of the Party Wall Act).*
>
> *104. Secondly, any failure to serve the requisite notice before work started would amount to a breach of statutory duty which would allow a court to award damages representing the compensation that would have been awarded by the surveyors appointed under the Party Wall Act for any damage caused by the work that would have been avoided had the notice provisions of the Party Wall Act been complied with.*
>
> *105. Thirdly, Mr Sampla's rights to claim damages for negligence, nuisance, trespass or withdrawal of support are not affected by the Party Wall Act ...*

To be clear, the PWA 1996 does not contain arbitration provisions. It also does not apply retrospectively, unless parties agree otherwise. Also, please note the decision in *Power v Shah*,[184] where it was held that:

> *First, it is clear that any rights which the building owner has under s.2 (including the right allegedly in play in this case, to cut away a projecting chimney breast under s.2(2)(g)), were expressly subject to s.3, and the service of a notice in advance of the works in accordance with s.3(1). The building owner has no right to carry out any of the works*

180 [2013] EWCA Civ 235.
181 ibid. 4.
182 (n 165).
183 ibid. 103–105.
184 (n 2) 25.

> *listed in s.2(2) unless he has served a notice under s.3. The service of the notice in advance is mandatory ("**before** exercising any right ... the building owner **shall** serve ..."). The service of the notice is therefore fundamental to the whole structure of this part of the Act.*

The Court of Appeal therefore held that the mechanism under the PWA 1996 was only engaged if an appropriate party wall notice had been served by the building owner on the adjoining owner ('no notice, no Act'). The conclusion therefore was that nothing in the PWA 1996 allowed an adjoining owner unilaterally to trigger the section 10 mechanism if no such section 3 notice had been served on it by the building owner. Therefore, unless an appropriate party wall notice had been served, the adjoining owner has no entitlement to invoke the PWA 1996. It was further held that this, however, did not leave the adjoining owner without a remedy as it could bring claims in trespass, nuisance and negligence, and seek an injunction if necessary. As noted earlier, as a practical solution, by way of contract parties could agree to invoke the provisions of the PWA 1996 retrospectively even where the building owner has not triggered the mechanism under the PWA 1996 with the relevant notice to begin with (i.e. works have started and/or been completed without the service of an appropriate party wall notice). To be clear, in such a situation, the PWA 1996 does not actually apply. The parties simply voluntarily follow the dispute resolution structure under the PWA 1996. This would allow parties to have recourse to a surveyor in the first instance rather than having to go to court if a dispute ensued resulting in waste of time, money and emotion. Should this strategy fail, parties can decide on an another agreed ADR avenue instead. If parties end up in court, they can decide at any point in time to suspend court proceedings in favour of settlement negotiation, for example, by way of mediation or other ADR methods. If such a settlement is reached, court proceedings can be formally terminated.

4.5.11 Contribution and Causation

A building owner's liability to pay damages for breaching the PWA 1996 can be the basis for assessing contribution under the Civil Liability (Contribution) Act 1978 from a contractor in a scenario where the contractor's works, which were carried out on the building owner's behalf, contributed to the damage due to the contractor's negligence.[185] The liability for damage caused to an adjoining owner's property rests with the building owner carrying out the works to the party wall. The building owner remains responsible for its contractor's works as the building owner's responsibility for the works to the party wall is non-delegable. However, the building owner may separately be able to recover (some) costs from the contractor if the damage was caused by the contractor's negligence.

Causation has been discussed in *Roadrunner Properties Limited v Dean,*[186] where the Court of Appeal has given guidance on how a court should scrutinise expert evidence with regard to causation of damage to property in a situation where the presumed cause is work

185 *Crowley t/a Crowley Civil Engineers v Rushmoor Borough Council* (n 165).
186 (n 164).

to a party wall where the building owner failed to serve notice on the adjoining owner under the PWA 1996. The Court of Appeal noted that:[187]

> ... *a court should be prepared to take a reasonably robust approach to causation. If it can be shown that the damage which has occurred is the sort of damage which one might expect to occur from the nature of the works that have been carried out, the court must recognise that the inability to provide any greater proof of the necessary causative link is an inability which results from the building owner's failure to comply with its statutory obligations. In those circumstances, as it seems to me, the court should be slow to accept hypothetical and theoretical reasoning in relation to causation advanced by the building owner after the event. It is within the building owner's power to ensure that proper evidence is or could be available; and if the conduct of the building owner has chosen to deny the adjoining owner the opportunity to obtain evidence, then the court should be slow to accept ex post facto and hypothetical reasoning and theory. The essential requirement, of course, is that the claimant proves the causal link which he or it asserts; but, as I have said, if there is material from which such a causal link can properly be established, I think a court, in those circumstances, should be slow to discard common sense in favour of expert hypothesis.*

4.5.12 Failure to Agree on a Surveyor and Acting Beyond the Realm of the Party Wall etc. Act 1996

Where either of the parties fails to agree to the appointment of a party wall surveyor and does not appoint its own party wall surveyor, the other party can appoint such a surveyor on the defaulting party's behalf under section 10(4) of the PWA 1996.

According to section 16 of the PWA 1996, it is an offence if the adjoining owner refuses a person access to its land to carry out works or obstructs a person from carrying out works (if the building owner has given the relevant notice of the proposed entry and is entitled to carry out the relevant works). In such a case, the building owner and its agents may break open with police assistance any fences or doors that prevent it from exercising its right of access to the adjoining land to carry out works according to section 8(2) of the PWA 1996. However, oversailing cranes or scaffolding over the adjoining owner's land and buildings are likely to constitute an actionable trespass.

If the building owner acts contrary to the party wall award or the rights it has under the PWA 1996 and causes loss and damage, it may be sued for damages in trespass or nuisance. It would also be liable for breach of statutory duty if it acts beyond the rights granted under the PWA 1996 and the party wall award. The adjoining owner's remedy then lies with the courts, and the remedy is damages (or an injunction might be sought where appropriate).[188] Under the PWA 1996, it is not possible for a party wall surveyor to retrospectively issue an award to authorise such works. If the building owner fails to complete the works or has

187 ibid. 29.

188 *Udal and another v Dutton and another* (n 169).

left the building in an unsafe and dangerous condition, the adjoining owner should raise the issue with the local authority building control department.[189]

4.5.13 Buying a Property Where a Party Wall Award Has Already Been Made

When buying a property where a party wall award has already been made, the buyer should take certain precautions:[190]

a) Successors in title – Some argue that party wall awards are personal to the original parties.[191] It should be considered how its rights under the PWA 1996 are affected. A buyer of the building owner's interest with the benefit of a contract for purchase is an 'owner' under section 20 of the PWA 1996 and therefore can serve a party wall notice prior to the completion of the transfer. A buyer without the benefit of a contract for purchase will have to wait until contract completion before it can serve a new party wall notice and go through the same procedures as the seller. However, save for circumstances where there is simultaneous exchange and completion, one cannot complete unless there is a contract in place first. To avoid any delays if the works continue after the sale, the buyer can give notice jointly with the current building owner meaning that the party wall award can be made with all relevant parties and take into account the current and future identities of the building owner (however this may not always be feasible). It should be added, however, that it is not entirely clear that the expression 'any adjoining owner' in section 7(2) of the PWA 1996 (in relation to compensation) includes successors in title to the original adjoining owner, although it may do so, as indicated by Nicholas Isaac KC.[192]

> *Section 7 is the most significant as to the rights of successors in title to the adjoining owner on whom notice was served. The requirement is that the building owner not exercise any right in such a manner or at such time as to cause unnecessary inconvenience to any adjoining owner. This gives scope to an adjoining successor in title who buys during the works by the building owner to sue for compensation. Similarly, an adjoining owner who buys after works have been completed, and discovers damage, appears to be able to seek compensation under section 7(2).*

However, the word 'any' in relation to 'adjoining owner' could equally mean merely any adjoining owner *at the time notices are served* regarding the building owner's works, i.e. since it is common for multiple adjoining owners to be served with notices by a single building owner in relation to the same works.

---

189 DCLG's Explanatory Booklet (n 26) 39.

190 Reid (n 77).

191 Some may favour the view that rights and liabilities under an award (or an agreement between owners in lieu of an award) comprise choses in action which may be transferable to successor owners. However, one has to consider *Observatory Hill Ltd v Camtel Investments SA* [1997] 1 E.G.L.R. 140, where it was decided that an adjoining owner's rights under an award could not be registered as a caution as they were not interested in land.

192 Isaac (n 23), 15–65.

b) Recent work – Where there was more recent work, the buyer should review the award, supplemental awards and other correspondence from the party wall surveyors to locate any damage or other problems. Enquiries of the seller should always be made.
c) Potential liability for breaches of the PWA 1996 – It has to be evaluated where such liability may rest after the sale. If the building owner's works go beyond the realm of the PWA 1996 and the adjoining owner suffers financial or physical damage as a result, the common law principles of nuisance, trespass and breach of statutory duty become relevant. The distinguishing factor is whether the issue giving rise to the loss was a one-off or is an ongoing tort.
d) The adjoining owner has consented to the works – Buyers of either building or adjoining owner's interests should look at the distribution of payments (for example, compensation or contribution towards expenses), which may arise after the sale but out of party wall procedures which started during the seller's time as owner.[193]
    i) Buying from the building owner – If the proposed works fall within section 2 of the PWA 1996, no notice by the building owner nor award is needed if the adjoining owner consents to the works in writing under section 3(3) of the PWA 1996. The adjoining owner's consent is likely to be personal to the original building owner. A buyer can prepare and serve a fresh notice on the adjoining owner once the buyer is contracted to buy or bought the building owner's interest.
    ii) Buying from the adjoining owner – An agreement should be reached with the seller that any compensation due to the adjoining owner under section 7 of the PWA 1996 will be apportioned appropriately. Further, any liability as to the vesting of the works in the building owner under section 14(2) of the PWA 1996 should be considered as well as contingent claw back provisions in sections 1(3)(b) and 11(11) of the PWA 1996.

4.5.14 Party Wall Dispute Resolution Procedure Under the Party Wall etc. Act 1996 – Conclusion

The dispute resolution process under the PWA 1996 is an important part of this book. While this book promotes other mechanisms as well, as set out further below, one must not forget about the PWA 1996 dispute resolution mechanism. This is far from perfect and is further clarified as well as scrutinised in case law. The dispute resolution process under the PWA 1996 is limited in its scope. However, it is an important and valid mechanism.

The level of detail of how disputes are addressed under the PWA 1996 shows both the reasoning and a decision tree how parties should behave under the PWA 1996 to solve party wall disputes and to benefit from the protections of the PWA 1996. While this is helpful to understand the rationale behind the PWA 1996 dispute resolution mechanism, it is not the only dispute resolution avenue parties can follow. The ADR options set out below can sometimes offer more pragmatic, emotionally sensitive, holistic (capable of covering other areas of law and fact) and cost-/time-effective solutions if the parties are willing to be open to them. One has to have due regard to the specific circumstances of each case in order to

193 *Selby v Whitbread & Co* (n 6).

make the best choice (having taken appropriate legal advice) in terms of which dispute resolution avenue is best for the disputants.

4.6 Other Forms of Dispute Resolution in the Context of Party Wall Disputes

4.6.1 Introduction

Following on from the above section and the party wall dispute resolution mechanism under the PWA 1996, let us take a step in a different direction, namely what other dispute resolution options there are for parties to party wall disputes. The party wall dispute resolution mechanism under the PWA 1996 is very useful, and other avenues in this book are provided to cover any gaps and/or offer solutions that suit parties better in certain situations. By considering and actioning alternative dispute avenues in situations where the PWA 1996 does not apply, where an award has been appealed in court or where ADR is better suited to parties' specific situation by way of agreement, members of the public have the potential to save money and time in relation to party wall disputes, and courts' workload would also be alleviated.

The emotional burden of a party wall dispute was by analogy well summarised in *Alan Wibberley Building Ltd v Insley*, a House of Lords decision, where Lord Hoffmann stated:[194]

> *Boundary disputes are a particularly painful form of litigation. Feelings run high and disproportionate amounts of money are spent. Claims to small and valueless pieces of land are pressed with the zeal of Fortinbras's army. It is therefore important that the law on boundaries should be as clear as possible.*

The above statement summarises the courts' view of neighbourly disputes in general. While the judicial attitude may continue in this direction, the fact remains that neighbourly disputes cannot be eradicated. Party wall disputes are a subcategory of neighbourly disputes and since they cannot be entirely prevented, there are ways to manage such disputes more efficiently.

4.6.2 The Protocol

Senior members of the legal and surveying professions have issued a protocol for disputes between neighbours on the location of their boundary (**Protocol**).[195] It is aimed at both residential and commercial landowners and occupiers, and the primary function of the Protocol is to guide parties through boundary disputes. However, parties can adopt the provisions of the Protocol without litigation being afoot. The Protocol attempts to set out a

194 [1999] 1 W.L.R. 894, 895.
195 S Tozer, G Fetherstonhaugh QC, J Karas QC, N Cheffings and M Ditchburn, 'Protocol for disputes between neighbours about the location of their boundary (The Boundary Disputes Protocol)' (2017), https://www.propertyprotocols.co.uk/the-boundary-disputes-protocol.

structured process for boundary disputes where parties exchange information in line with an agreed timeline, ideally leading to a swift dispute resolution minimising the cost, time and negative emotional impact involved. While the Protocol is not strictly a form of ADR, it enables parties to be provided with the appropriate information/documents by the other side, and to have opportunities to engage in mediation or other forms of ADR as part and parcel of it. In a sense, the Protocol might be a more effective method of resolving (boundary) disputes than merely launching into a mediation or 'without prejudice' negotiation uninformed of the other side's position and having seen none of their evidence.

4.6.3 The Party Wall etc. Act 1996 Dispute Resolution Process and Courts

When it comes to party walls, the PWA 1996 regulates the dispute resolution process whereby a party wall surveyor or surveyors issue a party wall award deciding the matters between the parties as long as the building owner triggers the dispute resolution mechanism under the PWA 1996 by serving an appropriate party wall notice.

Section 10 of the PWA 1996 applies to the resolution of disputes in relation to party walls and sets out a basis for the purposes of dispute resolution as to party walls, boundary walls and excavations located in the proximity of neighbouring buildings.[196]

As noted above, a party wall award can be appealed. However, there is a difference between a valid award, with which the appellant is not content, and an award that is defective and therefore null and void with no effect, which therefore requires no appeal. The assumption with a defective award should then be that there is no right to perform the works in the award or to pay any awarded costs. When a valid award is appealed, this can lead to either a complete rehearing where the judge will consider all facts afresh and usually on the basis of new evidence or simply by reviewing the award based on the tribunal of surveyors' reasoning and the material before them. The judge can modify or rescind the award. Costs usually follow the event in appeals, although it is for the court to decide what 'success' means in the context of the particular appeal before it.[197]

Section 10(17) of the PWA 1996 provides for appeals against party wall awards, which are heard in the county court. However, should a party not be content with the county court decision, the only possible appeal is to the Court of Appeal, which must itself give permission for such appeal. This is because the appeal from a county court decision related to a party wall award is a second appeal for the purposes of the CPR.[198] The Court of Appeal rarely grants such permission.

> *(2) The Court of Appeal will not give permission unless it considers that–*
> *(a) the appeal would–*
> *(i) have a real prospect of success and*
> *(ii) raise an important point of principle or practice or*
> *(b) there is some other compelling reason for the Court of Appeal to hear it.*[199]

196 Ministry of Housing, Communities & Local Government, 'Guidance – Preventing and resolving disputes in relation to party walls' (2016), https://www.gov.uk/guidance/party-wall-etc-act-1996-guidance.
197 *Taylor v Jones* [2024] EWCA Civ 170.
198 *Gray v Elite Town Management Ltd* (n 135).
199 The White Book Service 2024 (n 126), CPR Part 52.7(2).

Parties need to be aware that they may find themselves in a situation where the Court of Appeal does not give permission to appeal a county court judgment and the dispute therefore cannot be appealed further. An exception to this rule, however, is where the validity of an award is in question and where the county court finds that the award is not valid. Where a party wishes to appeal a finding of this kind, such an appeal is a first appeal. It is open to the county court judge to give permission to appeal (although permission may also be given by the High Court). The more important distinction is that the test for obtaining permission is 'real prospect of success' rather than the additional 'public interest' test applicable to second appeals.

In line with the general rule that appeals do not operate as a stay of the decision in the court below, while the PWA 1996 does not indicate whether the appealed decision is binding but not final or whether it is suspended, the logical assessment is that the appealed decision is binding but not final. This is because, for example, it would be unconscionable for an adjoining owner to delay the enforcement of an award by filing an appellant's notice. The building owner could incur significant loss, such as contractual delay costs or funding expenses. This would be particularly unfair where the appeal lacks merit. The adjoining owner has the option to apply for an interim injunction if it does not wish to comply with the award and to prevent its consequences while the appeal proceeds. At the same time, where an appellant wishes to apply for an interim injunction, it must generally provide the court with a cross-undertaking in damages. Such an undertaking is to cover losses if it is concluded that the injunction should not have been granted – generally where the appeal ultimately fails. Without the undertaking, the appellant's application for an interim injunction is likely to be refused by the court. If it is found that the application for the interim injunction lacks merit and the appellant becomes liable under the cross-undertaking in damages, there will be an enquiry determining how much should actually be paid. The enquiry can last months,[200] amounts to a new claim and is determined based on evidence.

One point worth mentioning from a strategic point of view is that 14 days is a very short time for an appeal under section 10(17) of the PWA 1996. While a party considering an appeal may wish to mediate, it should therefore keep in mind that mediation may fail and that once the 14 days under section 10(17) of the PWA 1996 have passed, an appeal will not be possible. The mediation process itself is likely to take more than 14 days. To safeguard its position, a party contemplating an appeal should appeal in time even if it wishes to mediate. In this way, a failed mediation will not cause such party to miss its deadline to appeal the award in county court. The court proceedings can then be stayed to enable mediation to take place. Should the mediation be successful, court proceedings can subsequently be formally terminated.

4.6.4 Parties Applying the Party Wall etc. Act 1996 Dispute Resolution Mechanism Voluntarily by Agreement, Where the Building Owner Failed to Issue the Relevant Notice Under the Party Wall etc. Act 1996

As mentioned earlier, and for completeness repeated here, where the building owner failed to serve the relevant notice under the PWA 1996 (i.e. the PWA 1996 does not apply to the relevant works as a result), as a practical solution parties could agree by way of contract to voluntarily follow the provisions of the PWA 1996 retrospectively (i.e. where works have

200 *Gray v Elite Town Management Ltd* (n 135).

started and/or been completed without the service of an appropriate party wall notice).[201] To be clear, in such a situation, the PWA 1996 does not actually apply, the parties simply voluntarily follow the dispute resolution structure under the PWA 1996. This would allow parties to have recourse to a surveyor in the first instance rather than having to go to court if a dispute ensues resulting in waste of time, money and emotion. Should this strategy fail, parties can decide on an agreed ADR avenue instead. If parties end up in court, they can decide at any point in time to suspend court proceedings in favour of settlement negotiation, for example, by way of mediation or other ADR methods. If such a settlement is reached, court proceedings can be formally terminated.

4.6.5 Alternative Dispute Resolution

Surveyors investigate the dispute and communicate their decision in a party wall award as discussed above. While this is often a practical way forward, it is not the only way to resolve a dispute in relation to party walls and the abundance of case law discussed in this book shows that. However, while it is possible to litigate a dispute, there are other methods to resolve disputes that may prove to be effective alternatives in the right circumstances where the parties do not want to resort to court.

Disputes on neighbourly matters have a tendency to escalate relatively quickly, which is why ADR avenues provide a good mechanism to resolve a dispute as early as possible resulting in saving both time and cost.[202] The main benefit of a negotiated settlement is that parties have the freedom to accept the terms of the settlement, influence those terms and, should the terms prove unacceptable to them, walk away from the settlement discussions. In addition, negotiated settlement saves cost and risk connected with a court process and gives the parties the option to keep the terms of the settlement confidential. The relationship between the parties to a negotiated settlement is more likely to 'survive' the dispute than if the parties go to court. This is particularly valuable in relation to neighbouring residential property owners, and also developers and contractors regarding construction sites as to new builds. If a relationship breaks down, it puts substantial strain on neighbours to the extent where one or both may sell up. Similarly, where there is a new development in the process of being built, preserving the relationship means a potentially smoother process of completing such a new development. A judgment as well as a settlement agreement is enforceable in court. However, parties may find a settlement agreement they reached themselves more palatable by comparison to a court judgment or a party wall ward.

ADR is growing in popularity, often providing a faster and cheaper way to achieve an understanding between parties. It also provides an avenue where parties may be better able to preserve a civil relationship as neighbours.

4.6.6 Courts' Support of Alternative Dispute Resolution

Parties can enter into settlement discussions and settle at any point, including before or during litigation (with the potential to apply for a stay of proceedings for the duration of

201 *Peck v Sims* CH-2022-000132 – If the building owner does not serve a notice and the adjoining owner would like to ensure the PWA 1996 will apply, the adjoining owner has to seek an injunction or agreement to restrain works pending notice and award (before works) or seek agreement to proceed as if notices had been served and to treat a retrospective award as valid (after works).
202 ibid. 74.

such settlement negotiations). Courts encourage amicable dispute resolution between parties and, where parties refuse ADR without a reasonable justification, this can result in court-imposed cost sanctions on either or both parties.[203]

However, it is not only the outright refusal to participate in ADR that is taken into account by courts. It is also the conduct of the parties during the ADR process itself that courts will consider where such conduct is viewed to be unreasonable. For example, in *Carleton (Earl of Malmesbury) v Strutt & Parker (A Partnership)*,[204] it was held that, where a party behaves in an unreasonable way during a mediation process, this equates to an unreasonable refusal to participate in mediation. In that case, this meant that the claimant's costs were reduced by 20%. However, the court was able to reach this decision because both parties to the dispute waived privilege covering the mediation. In most cases, parties will not be willing to waive privilege regarding mediation or settlement negotiations in general, and therefore the effect of this decision is somewhat limited.[205]

Parties can decide to mark relevant correspondence as 'without prejudice' (**WP**). The WP rule will generally (with exceptions) prevent statements made in a genuine attempt to settle an existing dispute (verbally or in writing) from being put before the court as evidence of admissions against the interest of the party that made them. As a result, where the WP rule applies, such statements are in general not admissible either in the substantive dispute or on questions of costs.[206] However, the situation is different where the statements are 'without prejudice save as to costs' (**WPSATC**). In such a case, the statements are not admissible in the substantive dispute, but they are admissible on questions of costs. When parties deal with issues on costs, a party may want to refer to the other parties' conduct/approach during the negotiations. Therefore, if correspondence is marked as WPSATC, once substantive issues of the matter are determined and judgment given, parties may refer to the WPSATC correspondence and decide on costs issues.

The reason why parties are usually unlikely to waive privilege in relation to mediation or settlement negotiation is because, when a party puts forward a sum as part of a settlement offer, this could sway the court when making a decision on financial sums awarded.

Where a party can show that it proactively sought ADR avenue(s) and the other party refused to co-operate, the court is required to take this into account when awarding costs. Courts are required to investigate what attempts the parties have made to resolve their dispute without continuing with court proceedings. As noted earlier, failure to engage with ADR can have significant consequences as to costs. Parties should protect its costs' position with one or more suitable CPR Part 36 offers.

Courts will take into consideration where the building owner fails to serve a party wall notice where the PWA 1996 applies, therefore breaching a statutory duty. If the parties agree on an ADR avenue (for example, mediation or expert determination), and one or both of the parties fail(s) to engage with such agreed ADR avenue prior going to court, this is likely to be viewed negatively by courts.

---

203 *Longstaff International Ltd v Evans (Costs)* [2005] EWHC 4 (Ch).
204 [2008] EWHC 424 (QB).
205 A Walsh, *A Practical Guide to Neighbour Disputes and the Law* (Law Brief Publishing, 2020) 73–74.
206 *Reed Executive Plc v Reed Business Information Ltd* [2004] EWCA Civ 887.

Where ADR fails and parties proceed to court, they need to follow the relatively rigid court procedure, which is likely to result in a substantial loss of time and money for the parties involved. ADR avenues can be much more attractive to the parties involved in a dispute. Further, courts, which are under financial pressure, encourage settlement at all stages of proceedings. Court proceedings provide parties to a party wall dispute with much less control and flexibility. In addition, and in general, court proceedings are public, so if the parties would like to keep the details of the dispute confidential, ADR may serve them better. Considering the relatively limited value in most party wall disputes, parties should consider how practical it is for them to go to court, where legal and other fees have the potential to rapidly become disproportionately high compared to the value of the subject matter in dispute. Unfortunately, and as mentioned earlier, due to such disputes often being of a highly emotional nature, parties sometimes resort to court proceedings ignoring the likely costs and time involved. While this is not ideal, and also not completely preventable, hope lies in ensuring that the parties are aware of the ADR avenues available to them both before and during court proceedings. Indeed, proceedings can be and often are suspended in order for the parties to be able to consider and discuss settlement options (and can be formally terminated, should a settlement be reached).

4.6.7 Methods of Alternative Dispute Resolution

The various methods of resolving disputes without recourse to litigation relate to ADR, which is an overarching term relating to a number of dispute resolution methods including mediation or expert determination.

As for arbitration, the Chartered Institute of Arbitrators defines arbitration as:[207]

> *... a private process, where a neutral arbitrator makes a legally-binding decision on the basis of the evidence provided by the parties. The arbitrator is empowered to do this by mutual consent of the parties, usually through a written agreement to arbitrate. An arbitration agreement can be incorporated into an underlying commercial contract, such as for goods, services, or purchase and sales.*
>
> *The parties may appoint the arbitrator of their choice, either directly or via an appointing authority. The arbitrator evaluates the evidence provided by the parties and then presents their decision in a written 'award'. The award is final, binding and enforceable and will – like a court judgment — usually find that one party is more right than the other. Arbitral awards can only be appealed in national courts in exceptional circumstances.*

It has been well established that the PWA 1996 dispute resolution process does not relate to arbitration or statutory arbitration[208] (even though section 10 of the PWA 1996 refers to an 'award'). To be able to have an arbitration award, there must be a binding arbitration

207 CIArb, 'What is dispute resolution?', https://www.ciarb.org/dispute-services/what-is-dispute-resolution/.

208 N Isaac, Party Wall & Boundaries Hub 'What can party wall surveyors learn from arbitration? (Part 1)' (2019), http://www.boundariesbook.co.uk/?p=985.

'agreement in writing' (as defined in section 5 of the Arbitration Act 1996, except in cases of statutory arbitration, see chapter V – HS2 Legislation) between the parties. Section 94(1) of the Arbitration Act 1996 applies provisions of its Part I to "*every arbitration under an enactment*," which it defines as a 'statutory arbitration'. However, a statute has to expressly refer to and create such arbitrations – for example, section 28 of the Agricultural Tenancies Act 1995, section 84 of the Agricultural Holdings Act 1986 or section 32 of the British Waterways Act 1995 refer to such arbitrations. The PWA 1996 does no such thing. The Arbitration Act 1996 does not apply to party wall awards falling under the PWA 1996. However, separately, parties are free to enter into an arbitration agreement to resolve their disputes. While arbitration is likely to amount to a more costly/longer process than mediation, the advantage is that the arbitrator makes the final and binding decision that can be then enforced in court, should need be. Arbitration is likely to be still cheaper/faster than going to court and even an arbitration can be brought to an end, should parties decide to settle during such a process. Parties can choose arbitration rules to govern the procedure of the arbitration and may have some say in selecting the arbitrator (or members of the arbitration tribunal). An advantage is that arbitration proceedings are usually confidential. The powers of the arbitrator (or arbitration tribunal) stem from the arbitration agreement, as supplemented by any applicable legislative provisions or arbitration rules. In general, the arbitrator (or arbitration tribunal) can decide on its own jurisdiction; however, such decision is subject to a court's review at the seat of arbitration. An arbitration award can be appealed in national courts only in exceptional circumstances. Arbitration in the context of party wall disputes is not particularly popular at the time of writing this book; however, it should be brought more to the forefront as a viable dispute resolution option with the right guidance from legal specialists. While an ADR method, it provides certainty where parties are not able to settle amongst themselves and therefore need an arbitrator to make a final and binding decision. Arbitration can be particularly helpful where the PWA 1996 is not engaged as the building owner failed to trigger the notice procedure. It can be also helpful where parties feel arbitration is more appropriate in their specific circumstances or where the party wall dispute is appealed in court and parties would prefer to resolve the dispute out of court.

Medi-arb is also available where parties would like to have certainty in securing a decision avoiding further delay, should mediation fail. In such a case, the mediator transforms into an arbitrator and will produce an arbitration award deciding the parties' dispute. However, one of the key problems with this process is that the mediator-turned-arbitrator may struggle not to be influenced by supposedly 'without prejudice' disclosures or suggestions made by the parties during the mediation. Parties may also not feel comfortable discussing their position openly at mediation stage due to this issue.

Early neutral evaluation is an option available under the Party Wall Mediation Scheme[209] where an experienced evaluator will consider parties' arguments in a potential appeal and give a view on the likely outcome if the appeal were to proceed to a court hearing.

---

209 Party Wall Mediation Scheme LLP, https://www.partywallmediation.co.uk; The Pyramus & Thisbe Society, the Faculty of Party Wall Surveyors and HHJ Edward Bailey set up jointly the Party Wall Mediation Scheme "*for the benefit of building owners and adjoining owners utilising the specialist skills and knowledge of party wall surveyors and lawyers practising in this field.*"

The aim of the evaluation is for the parties to be able to assess their prospects of success benefitting from the evaluator's informed opinion. Ideally this would then pave the path forward for reaching settlement.

Expert determination is a process where the parties agree to jointly appoint an expert, who then determines the issues that are in dispute between the parties. This is done on a binding basis. While generally expert determination has the potential to be significantly faster and cheaper over a county court claim, in complex cases expert determination can reach the cost and time investment of court proceedings.[210] It depends on the circumstances of each case.

Mediation is one of the most effective avenues when seeking to resolve a dispute amicably, especially where the parties struggle to communicate without a third party (a mediator). The cost and time involved is significantly lower for both parties than that of court proceedings. Mediation is relatively straightforward. The parties meet and an independent mediator encourages them to discuss the issues in dispute and reach a resolution, which is then recorded in writing so as to be enforceable in court as a contract. A mediator will often be a surveyor, barrister or solicitor with experience in party wall matters, but may also be very effective without such specialist knowledge. The mediation process is very flexible. Usually, the parties start by meeting in the same room – sometimes accompanied by their lawyers – where opening statements are made by the parties or on their behalf. Thereafter, or sometimes from the outset, the parties sit in separate rooms and the mediator goes to and from each of the parties as an intermediary, encouraging the parties to strip the issues to the main points and settle. This is particularly helpful where the parties are emotionally involved in the dispute as it helps de-escalate the situation by having an intermediary.

A settlement achieved at mediation provides more flexibility than a ruling of the court. As noted earlier, mediation is confidential (being covered usually by without prejudice privilege), and the parties benefit from the dispute and outcome not being on public display. The time and expense in relation to preparing for a mediation is limited compared to the much more laborious preparation for a trial. The only preparation usually involved in a mediation is for the parties or their lawyers to draft a position statement and to prepare the mediation bundle with relevant evidence each party relies on. The mediator/legal representatives will usually draft a binding agreement which can then be used to settle proceedings and be included in a Tomlin Order if the mediation is taking place during court proceedings (a type of consent order reciting that the parties have agreed terms of settlement and ordering that all further proceedings are stayed, but the parties are entitled to apply to court to enforce the agreed terms, should that become necessary, without having to commence new proceedings).

As can be seen above, mediations are not limited to being used only prior to court proceedings. On the contrary, mediations can take place at any stage before or during court proceedings. Strategically, however, if a mediation is to be started during court proceedings, parties will often time this to occur at an appropriate point in the proceedings. For example, after disclosure of documents or exchange of witness statements of fact. This is

210 Walsh (n 205) 76.

because, at these points in time, parties will have substantial information available to them to be able to assess their prospects of success in mediation. In terms of costs, parties generally each cover the costs of their own lawyers and half of the mediator's fees, at least in the first instance.[211]

It is recommended to consult a legal expert on the best ADR choice and strategy in each individual case.

4.6.8 Can Parties 'opt out' of the Party Wall etc. Act 1996?

The PWA 1996 is a statute, which does not explicitly permit to be 'opted out' of, and, as such, it cannot be 'opted out' of. However, according to recent case law, discussed further below, it is possible that the parties (building owner(s) and adjoining owner(s)) by agreement ensure that the dispute resolution procedure in section 10 of the PWA 1996 is not engaged. The PWA 1996 contains no provision preventing such an agreement.

In *Dillard v F & C Commercial Property Holdings Ltd*,[212] it was held that, where there is a potential dispute under the PWA 1996, the parties have the right to enter into a binding agreement resulting in the PWA 1996 not applying where they agree to pursue an ADR process.[213]

> *... the primary pointer to what they must have intended is the reference to "any dispute" arising under Clauses 7.5 and 10.4 being referable to expert determination. What the parties must be taken to have agreed is that all such disputes, whether otherwise verbally covered by the wording in Clause 11, would be referable exclusively under Clause 12 and not under the party wall dispute resolution procedure. (j) It is accepted rightly that parties may contractually opt out of the Act, as the parties have done here in part at least relating to the relief set out in Clauses 7 and 10 of the Deed.*[214]

While the author of this book takes issue with the phrase 'opt out of the PWA 1996', parties can ensure by agreement that the dispute resolution procedure in section 10 of the PWA 1996 does not apply and choose a form of ADR that suits them. In this case, the parties wished to proceed by way of expert determination.

> *The two or three surveyor procedure is limited in terms of jurisdiction by Section 10(10) to matters or disputes which are "connected with any work to which [the] Act relates" and which is in dispute between the building owner and the adjoining owner. The Clause 12 procedure goes very much wider at least in the context of disputes arising under Clauses 7.5 and 10.4 because they cover works which are not "connected with any work to which [the] Act relates."*[215]

---

211 ibid. 75.

212 (n 1).

213 N Isaac and M Hearsum, *The New Party Wall Casebook* (Property Publishing, 2019) 164–165.

214 *Dillard v F & C Commercial Property Holdings Ltd* (n 1) 17.

215 ibid. 17(f).

In *Dillard*,[216] the parties' contractual ADR clause applied to a much broader range of works, including those falling under the PWA 1996 and also those going beyond the scope of the PWA 1996. Rather than having separate dispute resolution procedures for party wall works falling under the PWA 1996 and works that do not fall under the PWA 1996, the agreement between the parties in *Dillard*[217] allowed them to ensure that the dispute resolution procedure in section 10 of the PWA 1996 did not apply, in favour of an ADR avenue applying to all of the works on the relevant site.

In other words, the effect of the agreement was not having different fragmented dispute resolution mechanisms in place running in parallel relating to different parts of the works on the relevant site. The developer in this case then ignored the agreement, which was one of the issues raised in court. The High Court decided that parties may contractually agree for the PWA 1996 not to be engaged.

In *Mohamed & Lahrie v Antino & Stevens*,[218] the claimant (an adjoining owner) sought and obtained an injunction preventing party wall surveyors from making an award where the parties had entered into an agreement which provided for how their past and future disputes should be determined.[219] The litigation prevented unwarranted interference by the surveyors with the parties' freedom to resolve their disputes in the way they saw fit.

For the claimants, Mr Isaac made three submissions, the first two of which are set out below and with which the court agreed:[220]

> (1) *The effect of the Consent Order is, explicitly, to resolve all disputes between the Takhars and Mohameds under the Act. The premise for the appointment of surveyors under section 10(1) of the Act is a dispute arising or being deemed to arise, and the sole purpose of those surveyors is to resolve that actual or deemed dispute. By agreeing that all disputes under the Act are resolved, the parties removed the very basis for the appointment of Mr Antino, or the selection of Mr Stevens, rendering them immediately redundant. Section 10(2) is not engaged because the Consent Order disengages section 10 completely;*
> (2) *It was open to the Mohameds and Takhars to contract out of the section 10 dispute resolution procedure, see Dillard v F & C Commercial Holdings Limited [2014] EWHC 1219 (QB), something the Consent Order explicitly did; . . .*

In relation to the second point, the court acknowledged that Mr Isaac relied on the comment of the judgment of Akenhead J. in *Dillard*, where he states:[221]

> *It is accepted rightly that the parties may contractually opt out of the Act, as the parties have done here in part at least relating to the relief set out in Clauses 7 and 10 of the Deed.*

---

216 ibid.
217 ibid.
218 (n 23).
219 N Isaac and M Hearsum (n 213) 262–263.
220 (n 23), 27.
221 (n 1) 17(j).

According to HHJ Bailey:[222]

> *The phrase contractually opt out of the act is, in essence, shorthand for the proposition, important in the present case, that the owners may by agreement ensure that the dispute resolution procedure provided by s 10 of the Act is not engaged.*

HHJ Bailey held:[223]

> *The argument that the 1996 Act requires the surveyors to proceed to an award even where the parties have reached agreement would be wholly against the general principle that parties who are sui juris are free to make such agreements as they wish to make, provided that they are not illegal in nature. It is also against the clear policy of the CPR for the court either to restrain two parties from reaching an agreement on any subject matter which has been referred to party wall surveyors under the 1996 Act, or to hold unlawful and therefore unenforceable any agreement they do happen to make, simply on the basis that a reference has been made to the surveyors. Very clear statutory words would be required before the court would act in this way, and there are no such words in the 1996 Act.*
>
> …
>
> *It would wholly inappropriate for the court to step in and interfere with the parties' contractual arrangements. Courts today are enjoined to encourage private resolution of litigation disputes, and there is no obvious reason why a different approach should be adopted to disputes which come within the 1996 Act dispute resolution procedure.*
>
> *Accordingly, I conclude that to the extent that the Agreement embodied in the Consent Order covers matters in respect of which the party wall surveyors have been appointed, the Claimants are entitled to the relief they seek.*

The main point is that on the clear wording of the PWA 1996 there must be a dispute before section 10 is engaged and if the parties reach an agreement the effect of which is that there is no dispute, then section 10 cannot apply. Of course, parties may resolve a party wall dispute by consent provided they are *sui juris* and as long as such an agreement is not illegal. Accordingly, it is open to owners to follow section 10 voluntarily by agreement, for example, by agreeing to accept a late party structure notice as if served before work commenced. The PWA 1996 will however still not apply.

Takhar v Mohamed[224] concerned the same settlement agreement as featured in *Mohamed & Lahrie v Antino & Stevens*[225] but was between different parties. In *Mohamed & Lahrie v Antino & Stevens*,[226] the parties were the adjoining owners and two of the party wall surveyors, whereas in *Takhar v Mohamed*,[227] the dispute was between the building owners and adjoining owners.

222 *Mohamed & Lahrie v Antino & Stevens* (n 23) 28.
223 ibid. 36–38.
224 [2023] EWHC 2190 (KB).
225 (n 23).
226 ibid.
227 (n 224).

The parties in *Takhar v Mohamed*[228] entered into a settlement agreement scheduled to a Tomlin order (after mediation) successfully, ensuring the PWA 1996 is not engaged. At the same time, the agreement purported to retain a right of appeal to the county court regarding the agreed surveyor's determination of compensation. However, the effect of the settlement was to take the dispute out of the scope of the PWA 1996. Under section 15 of the County Courts Act 1984, the county court had jurisdiction over a dispute regarding quantum. However, it did not have jurisdiction in relation to the contractual terms. It was held in the High Court that the parties were not able to reach a contractual agreement in relation to the scope of the county court's jurisdiction. The 'liberty to apply' a provision in the Tomlin order could not be applied to grant jurisdiction to the county court for a right of appeal in a situation where the PWA 1996 did not apply.

Foster J. held:[229]

> *... the language of the Tomlin Contract is clear in excluding the effect of the 1996 Act (from which, see the case of Dillard v F & C Commercial Holdings Limited [2014] EWHC 1219 (QB), they were entitled to contract out), in respect of both current and future disputes.*

Again, the author takes issue with the phrase 'contracting out of the PWA 1996' for the same reasons as taking issue with the phrase 'opting out of the PWA 1996'. Note however, that having a mechanism under the PWA 1996 relating purely to party wall disputes and a separate mechanism for any other works not falling under the PWA 1996 may be desirable to parties in different cases. This is a very personal choice for the parties in each case that has its unique circumstances. In addition, cases that relate purely to party wall works falling under the PWA 1996 are likely to benefit from the dispute resolution mechanism under the PWA 1996. Parties should be aware that they have a choice, by way of agreement, to pursue an ADR avenue instead. Whether this is appropriate or not depends on the specific circumstances of each case and wishes of the parties. There are party wall surveyors who provide services under the PWA 1996 but also offer mediation services (i.e. the ability to act as a mediator). Alternatively, a mediator does not have to be a party wall surveyor.

Parties should seek appropriate legal advice early so that they are aware of all legal implications of the works (including awareness of which works fall under a party wall surveyor's jurisdiction and which do not) in their specific circumstances and dispute resolution options available to them so that they can make an informed choice, which dispute resolution avenue they feel is suitable to their specific scenario.

At the time of writing this book, the decisions in *Dillard*,[230] *Mohamed & Lahrie v Antino & Stevens*[231] and *Takhar v Mohamed*[232] stand.

Irrespective of case law, the fact remains that, if the building owner fails to serve an appropriate party wall notice, both the building owner and the adjoining owner will lose the protections of the PWA 1996. Such statutory breach by the building owner is likely to

228 ibid.
229 ibid. 61.
230 (n 1).
231 (n 23).
232 (n 224).

be viewed negatively by courts. If the PWA 1996 is not engaged, parties are then left with the option of going to court or ADR avenues to resolve their dispute out of court. ADR can, by agreement, be pursued at any time – before going to court or during court proceedings.

It is prudent that any party wall surveyor appointed under the PWA 1996 in a party wall dispute checks whether there is an agreement in place between the parties the result of which is that the PWA 1996 is not engaged and choosing an ADR mechanism. The best way to do this is to obtain written confirmation from the relevant parties to the dispute. This is because the last thing the surveyor would want is to issue an award, which, if appealed in county court (with access to the case law discussed in this chapter), would clash with an agreement between the disputants resulting in the PWA 1996 not being engaged, such agreement pre-dating the award.

As noted by Mr Isaac in relation to *Dillard*:[233]

> *The main significance of this case is a higher court's recognition of the right of a building owner and adjoining owner to "contract out" of the 1996 Act, i.e. to agree that some or all of the provisions of the 1996 Act should not apply to certain building projects or operations. Most commonly, as here, such agreements impose even higher standards on building owners than the default position which would apply under the Act. Surveyors appointed under the 1996 Act, however, need to check whether any such agreements exist, and whether and to what extent such agreements impinge upon or limit their normal jurisdiction under the Act.*

As suggested above, according to HHJ Bailey, parties cannot agree that "*some or all of the provisions of the 1996 Act should not apply*" but rather that parties are able to reach an agreement the effect of which is that PWA 1996 does not apply to their particular dispute in the first place. The distinction may be subtle, but it is important as it is not open to parties to contract out of statutory requirements which apply to them except where the wording of the statute specifically provides that they may do so.[234] The PWA 1996 makes no such provision.

Parties should be aware that, when they enter into such an agreement and fail to then pursue the agreed ADR avenue in the relevant contract in place of the PWA 1996 dispute resolution mechanism, if a party pursues a claim in court the other party has no recourse outside of court. This is unless the parties decide to agree separately on an ADR avenue, as ADR is always an option at any stage (before and during court proceedings). If in such a scenario one party decides to go straight to court, the other party can raise the argument that the claimant is in breach of contract as it has not adhered to the ADR avenue agreed to by the parties and apply for an injunction to suspend such court proceedings. Having said that, it is questionable how effective this would be as during such suspension it is not possible to force a party to agree to ADR and settle. It is more likely that a court would impose cost consequences for the party failing to pursue ADR in such a scenario instead. Of course, the party initially refusing to pursue ADR could change its mind during such suspension. This, again, depends on the circumstances of each case. Parties would be wise to seek legal advice when drafting such contractual clauses.

---

233 (n 1); N Isaac and M Hearsum (n 213) 166.

234 For an example of a statute, which specifically provides that parties may contract out of a statutory provision see section 1(2) Contracts (Rights of Third Parties) Act 1999.

However, the notice procedure and dispute resolution mechanism under the PWA 1996 is a good dispute resolution avenue, and the building owner should adhere to it (unless it and the adjoining owner for very specific reasons together agree on an ADR avenue instead) and serve the appropriate party wall notice in relation to works falling under the PWA 1996. The building owner should keep in mind, as noted earlier in this book, that not complying with the PWA 1996 is a statutory breach and if a dispute arises, courts will not view the building owner favourably if it failed to serve an appropriate party wall notice where it should have done so, which is then followed by a dispute arising (without the parties having agreed an ADR avenue). At the same time, placing the financial and time burden on the adjoining owner having to resort to court as the PWA 1996 does not apply is not fair on the adjoining owner. One could argue, to ensure the building owner is not in breach of its statutory duty, it should serve the relevant notice in relation to works to which the PWA 1996 applies, even where it has agreed with the adjoining owner(s) on an ADR avenue separately and while not planning to pursue the dispute resolution mechanism under the PWA 1996, should a relevant dispute arise and parties are not successful in agreeing a settlement.

The building owner must not think that, by not triggering the PWA 1996, it can somehow circumvent its responsibilities since, if the adjoining owner decides to go to court, the consequences for both parties in terms of wasted time, money and emotion can be more severe than those which would have followed from a party wall award.

Appointing a party wall surveyor(s) experienced in party wall disputes under the PWA 1996, where it has been invoked by service of a notice, provides a swift and cheap dispute resolution process compared to litigation. However, parties can neither predict the outcome of such process nor influence the party wall surveyor in that process. The contents of the award are entirely a matter for the party wall surveyor(s). One or both/all disputants may not be happy with the award. Parties will have no choice but to comply with the award whether they agree with it or not. If the relevant party does not agree with the award, it can appeal in the county court and if the parties find themselves in court (or prior) and would like to stop the proceedings, they can resort to ADR by agreement.

ADR is voluntary. The parties are more likely to be motivated to comply with a compromise that they reached voluntarily than what would be prescribed by a party wall award or a court judgment. ADR fulfils its purpose if the parties cooperate. If the parties are not willing to cooperate, then an award under the PWA 1996 and/or a court judgment is more likely to provide a solution the parties will have to adhere to.

If the parties agree in their contract on an ADR avenue, and the agreement has the effect that the PWA 1996 is not engaged, it would be difficult for a disputant to argue that an award under the PWA 1996 should be issued instead, especially if the building owner did not serve a party wall notice prior. The dispute resolution mechanism under both the PWA 1996 and ADR has its place, and parties should be careful and pragmatic in evaluating their dispute resolution options, preferably after having sought appropriate legal advice.

4.6.9 Critical Considerations

There are a number of critical considerations set out in this book that parties to party walls should evaluate to ensure that any dispute resolution process runs as smoothly as possible. It is advisable for the parties to keep a record of their negotiations as well as any settlement attempts, which will be considered potentially as mitigating factors by courts, should the situation escalate. Any agreement between the parties as to how to proceed with works on a party wall should be recorded in writing for future reference and potential use in court to prevent any inconsistencies, miscommunications and establish certainty of what the parties agreed. The value of recordkeeping must not be underestimated and is crucial in any kind of dispute resolution process and even more so in court. It is always wise for parties to shield themselves in this way as it is impossible to say whether a potential dispute could arise and whether it may or may not end up in court. Preventative behaviour of this nature ideally assists all parties involved, including the judge, mediator, surveyor(s), experts, witnesses and, most importantly, the actual parties to the relevant party wall disputes.

It can be invaluable to maintain a trail of evidence showing parties' offers, suggested alternatives and lack of cooperation or response from the neighbours. Developers should also maintain evidence of their advance notice of proposed works. Maintaining such conduct that the relevant party can show to the court is more likely to win the court's sympathy.

It is a good idea to consult a lawyer specialised in party wall disputes and a chartered building surveyor for advice before party wall works start (whether as an adjoining owner or a building owner), which can save money and time in the long run. This will ensure that the party/parties are well informed about their legal rights, obligations and risks/benefits of any planned action, can communicate before a dispute ensues, and make sure there is a record of any such communications and any evidence in relation to the relevant works.

Obtaining legal advice before starting a dispute or shortly after the other party has started the dispute is a commendable step so that the parties are fully informed about their options and have an idea of their likelihood of success and best strategy based on legislation, case law and practical aspects benefiting from legal advisers' prior experience and specialised knowledge. This will also help the parties to present the strongest arguments possible. Consulting a lawyer as a preventive measure can assist in managing or even avoiding a dispute.

Transparency, communication and keeping the relevant authorities informed is a good way to mitigate the likelihood of complaints and resulting disputes. It also provides a robust defence where a statutory nuisance prosecution is started nevertheless by a local authority for breach of an abatement notice. A developer can then defend its case by highlighting that it complied with the necessary requirements. Unfortunately, this would not be considered a defence to a complaint lodged by a party, other than a local authority.

The best way forward is transparency and clear communication between the parties before issues arise. Once the parties are embroiled in litigation, usually the communication, transparency and co-operation between the parties slow down, which can pave the path for inflexibility and further escalation.

An adjoining owner should keep in mind to check their household or other insurance policy if it has legal expenses cover (and its limitations) and prepare the first draft of its witness statement before going to court.

Parties should also ensure there is a clear fee agreement in place with any lawyer/party wall surveyor and collect all relevant documents into a bundle which is paginated and indexed before going to court.

Party wall disputes are often linked to heightened emotions considering the personal nature of the disputes and that the parties are likely to have to come across each other after the dispute conclusion on a long-term basis due to party walls being linked to ownership/occupation of neighbouring land. Considering the potential need for the relationships to continue, ADR can offer a more palatable avenue than courts. This is why mediation is becoming increasingly popular as it saves both money and time, and also leads to a more amicable settlement on terms usually more acceptable to the parties than a court ruling or even a party wall award.

4.7 Conclusion

This chapter brings together the different dispute resolution mechanisms and options parties to party wall disputes may wish to pursue depending on the circumstances of each individual case. One should assess and give consideration to the different dispute resolution avenues outside of court. Of course, if a party to the dispute is not happy with the outcome of an award under the PWA 1996, it can appeal in court (if the award is valid).

While the dispute resolution mechanism under the PWA 1996 is commendable and provides a good practical avenue outside of court, it is not perfect and it is also not the only party wall dispute resolution option available to parties. This book promotes the importance and benefits of both the PWA 1996 dispute resolution mechanism and ADR to assist parties to save money, time and stress surrounding party wall disputes.

While it is not always possible to resolve a dispute or settle out of court, the PWA 1996 dispute resolution mechanism and ADR avenues are excellent opportunities to resolve party wall disputes relatively quickly and at a lower cost than going to court. For example, in ADR, parties to a settlement agreement have much more power over the terms of the arrangement and are more likely to willingly comply than in a situation where terms are dictated by the party wall surveyor/court and may result in dissatisfaction of one or more (or all) parties to the dispute. The court is likely to consider the conduct of the parties to the dispute, which also includes a building owner failing to serve the appropriate party wall notice in relation to works to which the PWA 1996 applies (breach of statutory duty) or a party refusing to participate in ADR.

In essence, parties to a party wall dispute have a certain level of control as to how far they are willing to go, how much they are willing to spend and how much time they want to devote to a dispute.

A third party (for example, a mediator) may be helpful in situations where parties have reached a negotiation standstill due to heightened emotions. Having a third party to assist the negotiation can help focus on the facts in dispute and on the pragmatic solutions that can benefit the parties.

5

Other Statutory Rights Relevant To Party Walls

5.1 Introduction

While the PWA 1996 focuses on party walls, it is not the only statute that is relevant to party walls. An award under the PWA 1996 cannot serve for the purpose of authorising works that infringe on another piece of legislation. What this means is that, for example, while an award can authorise certain works to be done, if a consent or permission is required under another statute, the building owner carrying out the works is not absolved from seeking such consent or permission just because it holds an award authorising its works under the PWA 1996.

Section 7(5)(a) of the PWA 1996 is also clear on awards not dispensing with any need for statutory consent. An award cannot, for example, allow for building works to be carried out without having to comply with the relevant safety requirements. An award can also not serve for the purposes of allowing materials to be used that are not permitted by the relevant building regulations. However, an issue may arise where the award sets out work suggested by a building owner to be done in a modified way due to technical reasons. This will be an issue particularly where the modification may result in the need for a planning permission or listed building consent or even for building regulations to be relaxed (which is possible under the powers conferred by the Building Act 1984 on local authorities). The fact remains that surveyors are not prevented by the PWA 1996 from awarding works that will need a statutory consent, including, for example, a planning permission or listed building consent under the Town and Country Planning Act 1990 and the Planning (Listed Buildings and Conservation Areas) Act 1990. The PWA 1996 also has no impact on building regulations requirements or work on construction sites regulations.[235] There are a number of key pieces of legislation that interact/overlap with or even, to a limited extent, override the PWA 1996. These key pieces of legislation are therefore analysed in this chapter.

235 Bickford-Smith, Nicholls and Smith (n 22) 15.26.

Party Wall Disputes: The Party Wall etc. Act 1996 and Beyond – *Legal Coherence and Dispute Management*, First Edition. Dr Laura Lintott.

5.2 Access to Neighbouring Land Act 1992

5.2.1 Relevance of the Access to Neighbouring Land Act 1992 to Party Walls

The Access to Neighbouring Land Act 1992 (**ANLA 1992**) is included in this book for a simple reason: it applies where the PWA 1996 does not in the context of landowners' access to neighbouring property to carry out works that are reasonably necessary for preserving their own land, or part of it.

The ANLA 1992 came into force in January 1993. Although it does not mention party walls specifically, it has a limited relationship with the PWA 1996. The ANLA 1992 does not use the same terminology as the PWA 1996; however for the purposes of this book, references to the building owner and adjoining owner used in the PWA 1996 are also used in connection with the ANLA 1992.

Without having an easement over the neighbour's land or the consent of the adjoining owner, there appears to be a point of standstill if a building is on or near a boundary where the building owner needs to carry out works to the building for the purpose of which it needs to access the adjoining owner's land. This is an example of a situation where works cannot be carried out without access to such a building.[236]

The ANLA 1992 gives the building owner the right to obtain access to the adjoining owner's land for the purposes of carrying out works in limited circumstances where the adjoining owner refuses to consent to the building owner having access to its land and where there is no easement that could assist the building owner in accessing the land in question. An example of the adjoining owner's consent is a situation where the adjoining owner and the building owner enter into a scaffolding or crane oversail licence (discussed further below).

According to *Megarry & Wade*,[237] rights emanating from the PWA 1996 have a substantially wider scope than those enjoyed by landowners that need access to their neighbour's property for the purposes of carrying out works that are reasonably necessary to preserve their own land (all or in part). These rights provided by the ANLA 1992 will only be relied on by a landowner where the PWA 1996 does not apply.[238]

Where the adjoining owner refuses to grant access to its land to a building owner, the building owner has the right to make an application to the county court under CPR Part 8 for the purposes of obtaining an access order. If the court issues an access order, it needs to be satisfied that the works are reasonably necessary for the preservation of the whole or any part of the building owner's land and that the works cannot be carried out without access to the adjoining owner's land (or that carrying out such work would be substantially more difficult without the requisite access).

236 ibid. 15.23.
237 (n 8).
238 ibid. 29-052.

There are certain works that are automatically considered to be reasonably necessary in order for the land to be preserved. Such works are called 'basic preservation works' and include those listed in section 1(4) of the ANLA 1992:

> *... (a) the maintenance, repair or renewal of any part of a building or other structure comprised in, or situate on, the dominant land;*
> *(b) the clearance, repair or renewal of any drain, sewer, pipe or cable so comprised or situated;*
> *(c) the treatment, cutting back, felling, removal or replacement of any hedge, tree, shrub or other growing thing which is so comprised and which is, or is in danger of becoming, damaged, diseased, dangerous, insecurely rooted or dead;*
> *(d) the filling in, or clearance, of any ditch so comprised; ...*

However, a court will refuse to issue an access order where the works would interfere with the adjoining owner's right to enjoy its land. Alternatively, the court will also refuse to issue an access order where the works to a party wall would cause hardship to the adjoining owner.

Where the court grants an access order, this has to include details of the work that the court allows to be carried out by the building owner on the adjoining owner's land as well as the timeline as to when such works can be carried out, and compensation should there be any loss incurred by the adjoining owner. Depending on the individual circumstances of each case, the access order can also include some other terms, for example, payment to the adjoining owner of a fair and reasonable sum in exchange for the building owner's privilege of entering the adjoining owner's land with the exception where the works are to residential land. Such sum is calculated by reference to the financial advantage of the order to the applicant (building owner) and connected persons, and the level of inconvenience that the respondent (adjoining owner or other person) may suffer due to the entry.[239]

Although there may be some overlap between the PWA 1996 and the ANLA 1992, where the PWA 1996 applies, the building owner has to comply with its provisions. A building owner is not permitted to use an access order issued under the ANLA 1992, in order to sidestep the requirements under the PWA 1996. Access rights that can be acquired under the ANLA 1992 are not necessary as the PWA 1996 gives a building owner rights to enter the adjoining owner's land (section 8 of the PWA 1996). Where the PWA 1996 does not apply, the ANLA 1992 may be applied to give a building owner the right to access an adjoining owner's land to carry out such works. Putting it explicitly, the ANLA 1992 cannot be used to circumvent the PWA 1996.

During the passage of the Bill in relation to the PWA 1996, the Earl of Lytton noted:[240]

> *The Bill will have no effect on title. The wall owned by one party or another is not a party wall, nor does it cut across provisions for easements contained in titles. It is not designed to affect common law rights or support or conflict with other statutory*

239 Bickford-Smith, Nicholls and Smith (n 22) 15.23–15.24.
240 ibid. 15.1.

> *requirements ... the Bill dovetails in with the Access to Neighbouring Land Act. I am satisfied that there is no conflict with that. Statutory consents such as planning, listed buildings and Building Regulations will be unaffected ...*

However, in the case of *Dean v Walker*[241] it was argued that the PWA 1996 does not dovetail in with the ANLA 1992.[242] In this case, the defendant needed to repoint a free-standing external wall and obtained an order giving them access to claimant's property so that the defendant could carry out repairs to the wall. The wall marked the boundary between the claimant's and the defendant's properties. The claimant resisted the application under the ANLA 1992 noting that the ANLA 1992 "*applied only to works to the dominant land and that either the whole wall, or at least the neighbour's side of it, was part of the servient land.*"[243]

The claimant appealed the order that granted the defendant access to the claimant's property. Grounds for the appeal included the two main arguments set out below.

a) The access order authorised work to be done on part of a party wall which belonged to the claimant. The ANLA 1992 gave no authority for the order to be made.
b) The defendant did not own the part of the wall on which the work was authorised.

According to the claimant, the application for access was also not necessary because the claimant was always willing to agree for the work to be carried out.

The appeal was dismissed on the grounds set out below.

a) As the affected part of the party wall that needed repair works to be carried out belonged to the defendant, it was entitled to the order as the repairs were reasonably necessary and there was no evidence of the claimant's willingness to consent to the necessary access.
b) The Court of Appeal determined that the court at first instance was correct in that a wall could be a party wall in part and belong in another part to one of the joint owners separately. (This followed judgments in *Knight v Pursell*[244] and *Weston v Arnold*.)[245] The wall in this case was coextensive between the parties' properties and therefore belonged to the defendant. Thus, it was not necessary to determine whether the ANLA 1992 applied to works on a neighbour's land.[246]

The Court of Appeal concluded that the ANLA 1992's reference to works 'to land' is not limited to dominant land only. Therefore, it was not necessary in this case to determine ownership of the wall, which was split vertically.

The Court of Appeal's judgment shows that there is an overlap between the PWA 1996 and the ANLA 1992. The wall in the case was a party fence wall. Therefore today (at the time of the case, the PWA 1996 was not in force yet), the ANLA 1992 would not be necessary or sufficient for the building owner's purpose. An order under the ANLA 1992 cannot

241 (1996) 73 P. & C.R. 366, CA.
242 Bickford-Smith, Nicholls and Smith (n 22) 15.25.
243 ibid.
244 (1879) 11 Ch. D. 412.
245 (1872–73) L.R. 8 Ch. App. 1084.
246 *Dean v Walker* (n 241).

override the compulsory requirements of the PWA 1996. In addition, an award under the PWA 1996 now carries rights of entry allowing access without having to apply for an order under the ANLA 1992. It would therefore seem that, within the overlap between the PWA 1996 and the ANLA 1992, where the works are needed within the PWA 1996, the ANLA 1992 no longer applies. This limits the scope of the ANLA 1992 substantially.

However, there may be situations where the building owner will need to apply both the PWA 1996 and the ANLA 1992. This would be the case where the building owner needs access for repair or maintenance work to its own property not falling within the PWA 1996 (where the ANLA 1992 applies) as well as a party wall (where the PWA 1996 applies). This is because, where the purpose of works is not carried out pursuant to the PWA 1996, an award issued under the PWA 1996 does not authorise entry in relation to such works.[247]

For the purposes of this book, the ANLA 1992 is noted because of its interplay with the PWA 1996. However, its scope and application are fairly limited in this context but for the exception mentioned above regarding the building owner's need of access for work to its own property not falling within the PWA 1996.

5.2.2 Practical Considerations – Scaffolding and Crane Oversail

In relation to development sites where construction is ongoing, arranging legitimate access is crucial to be able to facilitate such construction efficiently, to maintain a property and to build near a boundary. For this purpose, developers can seek to enter into scaffold or crane oversail licence agreements with adjoining owners, which includes relevant property owners as well as leaseholders.

Where there are works that do not fall under the PWA 1996, access to the relevant property or party wall without permission from the adjoining owner constitutes trespass. Therefore, permission is required from the adjoining owner prior to accessing its land for the purpose of legitimate repairs.[248]

However, it is not possible to force the adjoining owner to enter into, for example, a scaffolding or crane oversail licence. Similarly, where there is a right of way over another property within the deeds, this does not automatically mean that the developer or building owner has the right to erect scaffolding or a crane over the adjoining owner's land.

Section 8 of the PWA 1996 (focussing on rights of entry) is likely to extend to erection of scaffolding as long as it is used for the purposes of works under the PWA 1996. However, section 8 of the PWA 1996 does not obviously entitle the building owner to oversail cranes above the adjoining owner's land, which would amount to actionable trespass. The main reason is because scaffolding is likely to be needed for the required access to perform works in relation to the party wall, whereas the same argument is less likely to apply to cranes. It is conceivable that a crane might be required in order to undertake notifiable work, and this would obviously entitle the building owner to access for that purpose. However, it would be very rare in practice, and it is unlikely that a crane would be required only for notifiable work.

To err on side of caution, it is advisable to enter into a scaffolding/crane oversail licence agreement and set out the exact nature of the scaffolding/crane oversail, the duration of the

247 Bickford-Smith, Nicholls and Smith (n 22) 15.25.

248 *John Trenberth Ltd v National Westminster Bank Ltd* (1980) 39 P. & C.R. 104.

scaffolding/crane being present and its purpose as well as any pecuniary advantage obtained by the adjoining owner under the licence.

In *London and Manchester Assurance Co v O & H Construction*,[249] a dispute arose between adjoining owners of areas of old London quays. The court held that the defendants unlawfully encroached on the claimants' property. This is because the defendants demolished a party wall without obtaining a consent and in its place built their own structures encroaching onto the claimants' land. The court made mandatory interlocutory orders so that the offending structures were to be removed. The defendants also built over part of a footpath that was on the claimants' part of the boundary. In addition, the defendants swung a crane over the claimants' land without having obtained a permission prior to doing so. In his judgment, Harman J. noted:[250]

> *I did, however, make an order restraining any trespass by the boom or jib of the crane over the plaintiffs' land. It is, of course, notorious that the use of a crane swinging round from the useful position upon a neighbouring site is liable to intrude into the airspace of other persons and thereby to commit a trespass. The matter has been the subject of a good deal of litigation. It is, in my view, beyond any possible question on the authorities and the law that a party is not entitled to swing his crane over neighbouring land without the consent of the neighbouring owner. Upon that basis, I accepted an undertaking, having said that I should impose an order. An undertaking was of course just as binding and even more effective than an order, because it did not need service and a penal notice endorsed upon it in order to bite. The purpose of the undertaking was to restrain any permission or causation of any part of the crane trespassing over the plaintiffs' land. As it seems to me, that undertaking which I accepted over the effective hearing of the motion can only properly now be continued, if necessary by order, over trial of this action. I have heard nothing from Mr. Ellis, who has been most eloquent in his attempt to make bricks without straw, which could conceivably justify the trespass by swinging the crane over the neighbouring site.*

It is irrelevant whether the crane is or is not carrying a load over the adjoining owner's land as the jib itself can constitute trespass (without a crane oversail licence, where the PWA 1996 does not apply and where there is no express right in the property deeds). This is based on the decision in *Anchor Brewhouse Developments Ltd v Berkley House (Docklands Developments) Ltd*,[251] where Scott J. held:

> *It is in my judgment well established that it is no answer to a claim for an injunction to restrain a trespass that the trespass does no harm to the plaintiff. Indeed, the very fact that no harm is done is a reason for rather than against the granting of an injunction: for if there is no damage done the damage recovered in the action will be nominal and if the injunction is refused the result will be no more nor less than a licence to continue the tort of trespass in return for a nominal payment.*[252]

249 (1990) 6 Const. L.J. 155.
250 ibid. 158.
251 (1988) 4 Const. L.J. 29, 33.
252 ibid. 33.

A developer, who needs to use a crane that will oversail over the adjoining owner's land, needs to communicate with the adjoining owner in order to secure an oversail crane licence agreement, usually for a pecuniary advantage. Again, the adjoining owner is in a stronger position during such related negotiations particularly if the developer's scheme depends on the adjoining owner's acquiescence, which may drive the pecuniary sum upwards. In legal terms, however, the licence will protect the developer and ensure that it is not committing trespass[253] onto the adjoining owner's land when the developer's crane is oversailing the adjoining owner's land.

5.3 Boundary Determination and Related Legislation in the Context of Party Walls

5.3.1 Introduction

Boundary determination is an important topic to be analysed in conjunction with party walls as these can demarcate boundaries between properties. The PWA 1996 refers to 'boundary walls', as section 1 of the PWA 1996 notes, that will have effect:

> *... where lands of different owners adjoin and (are not built on at the line of junction; or (b) are built on at the line of junction only to the extent of a boundary wall (not being a party fence wall or the external wall of a building), and either owner is about to build on any part of the line of junction.*

Further, section 2(1) of the PWA 1996 relates to:

> *... where lands of different owners adjoin and at the line of junction the said lands are built on or a boundary wall, being a party fence wall or the external wall of a building, has been erected.*

In light of the PWA 1996 referring to 'boundary walls', it needs to be clarified what a 'boundary' is and how it is determined, as well as its link to party walls. Considering the boundary of a property determines what rights and liabilities of a property owner or occupier will operate, it is important to have clarity as to how boundaries are established.

The characteristics of a boundary include those set out below.[254]

a) A boundary is a line that divides two contiguous parcels of land.[255]
b) A boundary may be physical, in that it is marked by a physical feature, which may be natural (such as a river)[256] or artificial (such as a wall).
c) A boundary is fixed by acts of owners that are proven, statutes/orders of relevant authorities having jurisdiction or by legal presumption (where there is no statute or order fixing the boundary).

---

253 *John Trenberth Ltd v National Westminster Bank Ltd* (n 248).
254 Halsbury's Laws of England Online (LexisNexis) 301.
255 *Stepney Corp v Gingell Son & Foskett Ltd* [1909] A.C. 245.
256 *Scratton v Brown* (1825) 4 B. & C. 485; *Bickett v Morris* (1866) L.R. 1 Sc. & Div. 47.

The first source of information are the title deeds establishing the boundary line. If part of the description is correct, that part can be read on its own and independently from an incorrect part, which can be rejected.[257] If there are several conflicting descriptions of the boundary line, the court must establish the parties' true intentions. Although this should be done from the written instrument, extrinsic evidence can be admitted if the description is too general, contradictory, uncertain or ambiguous to set out the correct boundary lines.[258]

Where the description and plan conflict:

a) where the plan is described to be 'for identification purposes only', the description is the prevailing source,[259] but where the description is not adequate the court may conclude that the plan determines the parties' intentions;[260]
b) where the description of the property includes the wording 'more particularly delineated or described on the plan', the plan will prevail;[261] and
c) where the deed describes the plan as being 'for identification purposes only' and further describes the property as being 'more particularly delineated on the plan', the two statements cancel each other out in which case neither of the plan or description dominates.[262] It is ultimately the grantor's responsibility to ensure that a plan is accurate as a deed is construed *contra proferentem* (against the interests of the grantor). An erroneous plan may be disregarded where the parcels clause is clear.[263] If there is a transfer or conveyance describing the land by reference to a plan of a very small scale, related circumstances need to be taken into consideration.[264] Where there is a conflict between the dimensions on a plan and measurements scaled from it, resolution comes from reference to extrinsic evidence.[265]

Determining where the boundary lies is crucial in cases where the party wall is positioned on such a boundary. Where the boundary is not clearly set out in the title deeds, there are a number of legal boundary presumptions that can assist. It is possible to submit evidence to rebut such presumptions. Examples of established legal presumptions include roadways, hedges and ditches, non-tidal rivers and streams, lakes or seashores. However, these cannot apply when section 60 of the Land Registration Act 2002 determines the boundary.[266]

---

257 *Llewellyn v Earl of Jersey* [1843] 11 M. & W. 183.
258 *Lyle v Richards* (1866) L.R. 1 H.L. 222.
259 *Hopgood v Brown* [1955] 1 W.L.R. 213; *Neilson v Poole* (1969) 20 P. & C.R. 909.
260 *Wigginton & Milner Ltd v Winster Engineering* Ltd [1978] 1 W.L.R. 1462; *Targett v Ferguson* (1996) 72 P. & C.R. 106.
261 *Eastwood v Ashton* [1915] A.C. 900.
262 *Neilson v Poole* (n 259).
263 *Maxted v Plymouth Corp* [1957] 3 WLUK 80.
264 *Gillon v Baxter* [2003] EWCA Civ 1591; *Cameron v Boggiano* [2012] EWCA Civ 157.
265 *Cook v JD Wetherspoon Plc* [2006] EWCA Civ 330.
266 D Cavill, S Wheeler, Professor M Dixon, *Ruoff & Roper, Registered conveyancing* (Sweet & Maxwell, 2021) H40C.012-019.

5.3.2 Registered Land Boundaries – General Boundary Rule

Ordnance Survey maps form the basis of Land Registry plans. However, such Ordnance Survey maps are subject to accuracy and plotting limitations. Land Registry title plans are created, based under the 'general boundaries rule' as the legal boundary is usually uncertain. This means that the exact boundary line remains undetermined by the Land Registry with the exception where an application is made for the exact boundary line to be ascertained, which eventually leads to this being added to the Land Registry plan. The Land Registration Rules 2003/1417 (**LRR 2003**) are used to set out the general boundary rule and the procedure for fixing boundaries, supporting the Land Registration Act 2002 (**LRA 2002**). Both came into force on 13 October 2003. As the Land Registry accepts there may be mistakes in the register, it is possible to make an application for the register to be altered based on accurate evidence.[267]

According to section 60(3) of the LRA 2002:

> *(1) The boundary of a registered estate as shown for the purposes of the register is a general boundary, unless shown as determined under this section.*
> *(2) A general boundary does not determine the exact line of the boundary.*
> *(3) Rules may make provision enabling or requiring the exact line of the boundary of a registered estate to be determined and may, in particular, make provision about–*
> *(a) the circumstances in which the exact line of a boundary may or must be determined,*
> *(b) how the exact line of a boundary may be determined,*
> *(c) procedure in relation to applications for determination and*
> *(d) the recording of the fact of determination in the register or the index maintained under section 68.*
> *(4) Rules under this section must provide for applications for determination to be made to the registrar.*

According to rule 118 of the LRR 2003, the exact line of the boundary is to be determined when the proprietor of a registered estate makes an application to the registrar, who, once satisfied, must give notice of the application to the owners of land that adjoin the boundary to be determined. The notice is not needed where an adjoining owner has entered into a written agreement setting out the exact boundary line or where a court order has been issued determining the boundary according to rule 119(2) of the LRR 2003. The recipient of a notice can object within the period ending at noon on the twentieth working day after the date of when the notice had been issued according to rule 119(3) of the LRR 2003 unless extended. According to rule 117 of the LRR 2003, the right to apply for a boundary to be determined also applies to a right to apply for only a part of a boundary to be determined. Rule 122 of the LRR 2003 allows the registrar to determine the exact line of a common boundary in certain circumstances.

Boundary fixing has historically been time consuming and expensive. It is a procedure that is rarely used as the applicant has to cover the costs of all the works. These difficulties

267 HM Land Registry Practice Guide 77, 'Altering the register by removing land from a title plan', https://www.gov.uk/government/publications/altering-the-register-by-removing-land-from-a-title-plan-pg77/practice-guide-77-altering-the-register-by-removing-land-from-a-title-plan.

remain under the LRA 2002 and are even more complex due to rule 119 of the LRR 2003 setting out the procedure on an application for the determination of the exact line of boundary. Considering, however, that fixing boundaries can prevent a squatter from acquiring the right to be registered as proprietor as a result of adverse possession, fixing boundaries may become more common.

The Land Registry aims to show the land and its boundaries as accurately as possible. Where a person believes that a legal presumption exists, has valid grounds for this and also believes that the red edging on a title plan should include additional land, they may apply for an alteration to be made to the register. Even so, the boundary will remain a general boundary. Parties may decide to enter into a boundary agreement.

There are two types of boundary agreements:

a) those that amount to an exchange of land; and
b) those where parties merely intend to demarcate an unclear boundary that is referred to in title documents.[268]

An agreement merely demarcating an unclear boundary may be informal.[269] A boundary agreement can also be inferred from the parties' conduct. These types of agreements are ideal in that they save time and cost for neighbouring property owners, can avoid a dispute as well as procedural steps related to the application process under the LRR 2003.

Boundary determination can be highly contentious, and the main principles applied to boundary disputes are set out in the decision in *Acco Properties Ltd v Severn*.[270] When it comes to party walls, boundary disputes are highly relevant, as further noted below, and particularly as party walls can be erected on boundary lines.

5.3.3 Law of Property Act 1925 and Party Walls

The Law of Property Act 1925 (**LPA 1925**) is relevant both to boundaries and to party walls. Section 38(1) of the LPA 1925 sets out that a boundary runs along the centre of a party wall:

> *(1) Where under a disposition or other arrangement which, if a holding in undivided shares had been permissible, would have created a tenancy in common, a wall or other structure is or is expressed to be made a party wall or structure, that structure shall be and remain severed vertically as between the respective owners, and the owner of each part shall have such rights to support and user over the rest of the structure as may be requisite for conferring rights corresponding to those which would have subsisted if a valid tenancy in common had been created.*

To add clarity as to the distinction between a 'party wall' and 'boundary wall', section 1(1) of the PWA 1996 is helpful in that it states:

> *(1) This section shall have effect where lands of different owners adjoin and-*
> *(a) are not built on at the line of junction; or*
> *(b) are built on at the line of junction only to the extent of a boundary wall (not being a party fence wall or the external wall of a building).*

268 *Neilson v Poole* (n 259).
269 *Acco Properties Ltd v Severn* [2011] EWHC 1362 (Ch); *Charalambous v Welding* [2009] EWCA Civ 1578.
270 (n 269).

In light of this, section 20 of the PWA 1996 needs to be highlighted as it defines the terminology 'party wall' as:

(a) A wall which forms part of a building and stands on lands of different owners to a greater extent than the projection of any artificially formed support on which the wall rests; and
(b) so much of a wall not being a wall referred to in paragraph (a) above as separates buildings belonging to different owners.

The three legislation quotations assist in assessing what amounts to a party wall and case law set out below further clarifies this.

In *Wellington Properties Ltd v Trustees of the Will of the Second Duke of Westminster*,[271] the main issue was whether the flank wall of 39 Headfort Place, London SW1X 7DE (**39HP**) adjoining the Halkin Street Garden was a party wall within the meaning of section 38(1) of the LPA 1925 and of section 20 of the PWA 1996. In turn, on appeal, the main issue was whether the judge was correct on the true construction of a lease transfer dated 25 April 1990 where Grosvenor transferred 39HP to Lady Anthea Peronelle Rees and where the claimant was the successor in title.[272]

Clause 3(b) of the transfer provided that it was agreed between the parties:

> *... that the walls and/or fences separating the property hereby transferred from adjoining properties are party walls and/or fences and shall be used maintained and repaired as such.*[273]

The court at first instance agreed with the claimant's argument that the clause meant that the walls separating 39HP from adjoining properties, including the garden wall, were party walls. The defendant argued that the words relate to 'adjoining properties' and therefore mean 'adjoining buildings', which the garden is not and so the garden wall should not be construed to be a party wall.

The High Court agreed on appeal with the defendant's interpretation of clause 3(b):[274]

> *... (ii) It is, of course, correct, that, where a terraced house is enfranchised, then the freehold acquired by the enfranchisor will only include half of each party wall, But the Garden Wall is not a wall between adjacent houses in a terrace, it is a wall between a house and a garden. On Wellington's construction, Lady Rees failed to acquire all of the house which she was entitled to. That is contrary to the expressed purpose of the Transfer. ...*
>
> *(iv) The plan shows the boundary between 39HP and Headfort Place as being the front wall, not the middle of the road. If that is correct, it would be surprising if that wall were a party wall. (The arrows indicate that it is not.) ...*

In the context of boundary issues, it is important to look at the LPA 1925 (and section 38(1) of the LPA 1925 in particular, in conjunction with sections 1 and 20

271 [2018] EWHC 3048 (Ch).
272 ibid.
273 ibid. 20.
274 ibid. 41.

of the PWA 1996) when assessing the existence of party walls. The turning point in this case was clause 3(b) of the transfer.

Another case, *Burlington Property Co Ltd v Odeon Theatres Ltd*,[275] shows a dispute between a building owner and an adjoining owner as to a party wall where section 39(5) of the LPA 1925 was referred to and relied on by the claimant (later appellant). For ease of reference, Diagram 5.1 provides an image showing the positioning of the party wall.

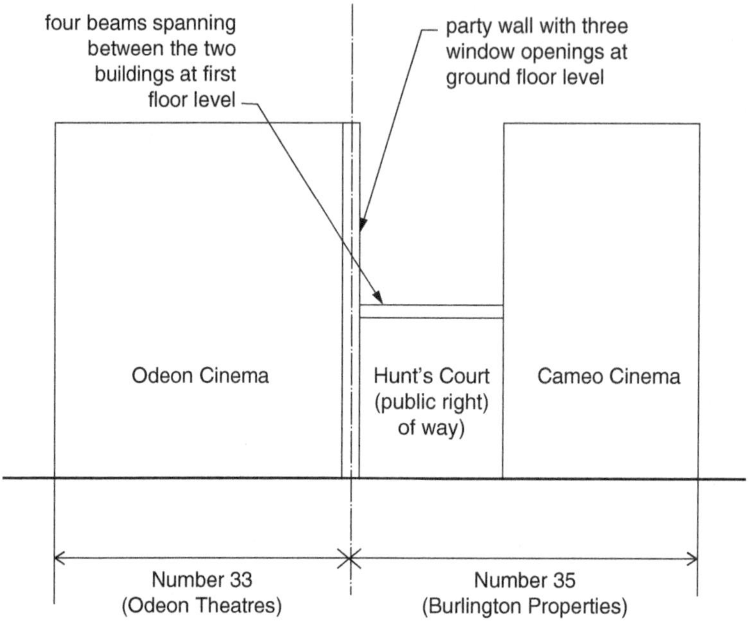

Diagram 5.1 *Burlington Property Co Ltd v Odeon Theatres Ltd.*[276]

The First Schedule, Part V of the LPA 1925 at paragraph 1 contains an important consideration in relation to boundary determination in the context of party walls:

> *Where, immediately before the commencement of this Act, a party wall or other party structure is held in undivided shares, the ownership thereof shall be deemed to be severed vertically as between the respective owners, and the owner of each part shall have such rights to support and of user over the rest of the structure as may be requisite for conferring rights corresponding to those subsisting at the commencement of this Act.*

The case relates to the taking down and rebuilding of a party wall (several storeys tall). At first instance it was held that the appointed surveyors had no jurisdiction to act as 'arbitrators' and to make an award. The main issue was whether the building owner (defendant)

275 [1939] 1 K.B. 633.

276 P Chynoweth, *The Party Wall Casebook* (Blackwell Publishing Ltd, 2007) 81.

was entitled to substitute existing windows in the ground storey portion of the party wall with a wider opening coming down to the ground level so that foot passengers could pass through them and from an adjoining public court. It was held:[277]

> *The judgment of the county court judge was wrong in so far as it affirmed the award of the arbitrators in determining that the respondents might rebuild the party wall with openings in it, or at all events with openings larger than the windows which it formerly contained. The wall is a party wall. It has all along been used for the separation of the adjoining premises now respectively vested in the appellants and respondents, and the evidence shows that these parties and their predecessors have been and are respectively entitled to one half of the wall and to the use of the wall for the support of their respective buildings. It is therefore a party wall within the definition of that expression in the Law of Property Act, 1925, s. 39, sub-s. 5, and First Schedule, Part V., clause 1, the London Building Act, 1930, s. 5, and Watson v. Gray. If the wall was originally held in undivided shares, then the provisions of the Act of 1925 sever it vertically and give to the appellants the side next to their premises with rights of support and user over the rest of the wall; so that in any case the appellants have these rights in the wall. In the absence of any provision or agreement to the contrary, the expression "party wall" means a solid and continuous party wall.*

The wall was considered to be a party wall. The Court of Appeal held that the defendant could only demolish and rebuild the party wall in accordance with substantially the same design as it was. Therefore, the defendant was not entitled to rebuild the party wall to a different design.

The topics of boundaries and party walls are intrinsically intertwined considering that party walls affect two or more parties who are landowners or occupiers. It is therefore important to carefully assess where exactly boundaries lie in relation to party walls and for this purpose to consider the title deeds establishing boundary lines, legal boundary presumptions and relevant legislation, including the LPA 1925 (which sets out where the boundary runs in relation to a party wall in section 38(1)), LRA 2002 (which sets out the general boundary rule in section 60), PWA 1996 (setting out what constitutes a party wall in section 20) and also consider the relevant case law. Therefore, where party walls are intertwined with the issue of boundary determination, it is important to consider all of the above aspects and not to focus only on the party wall in connection with the PWA 1996 in isolation. This is because a party wall's position and alignment with underlying boundary lines may determine the relevant parties' rights and obligations.

5.4 Property Boundaries (Resolution of Disputes) Bill [HL] 2019–2020

The very same Earl of Lytton, who introduced the PWA 1996, attempted to divert boundary disputes away from litigation and from tribunals with a similar mindset as that behind the PWA 1996 showing that property boundaries and related disputes are connected with party

---

277 *Burlington Property Company, Limited v Odeon Theatres, Limited* (n 275) 637.

wall disputes. He wanted the disputes to be dealt with by surveyors. As a result, the Property Boundaries (Resolution of Disputes) Bill [HL] 2019–20 (**Bill**) was introduced as a Private Member's Bill into the House of Lords on 13 July 2017 for the first reading. However, the new proposed legislation stalled. The Bill did not progress past the first reading and no date has been set for its second reading.

The reason, however, why the Bill needs to be mentioned in this book is because of its purpose, which is to set out a dispute resolution structure in relation to the location or placement of boundaries and private rights of way as to an estate in land title. This touches on party walls, and it makes sense why both the Bill and the PWA 1996 stem from the same mind-set.[278] In fact, the Bill was modelled on the PWA 1996. Both the PWA 1996 and the Bill aim to make the dispute resolution process more straightforward and faster for adjoining neighbours rather than going to court in relation to party walls and boundaries/rights of way, respectively, areas, which often overlap (as is also clear from Chapter VI covering easements and their relevance to party walls). Both the PWA 1996 and the Bill aim for surveyors to be appointed for such dispute resolution purposes.

In practice, boundary disputes, especially in residential context, arise rarely about the actual land. Disputes usually revolve around broader issues between the parties, with such issues being a sore spot for the parties for protracted periods of time. The level of animosity in boundary disputes is often inversely linked to the size of the land the dispute relates to. Once such a dispute starts, it is rare for the parties thereafter to be able to harmoniously live next to each other, especially after a court decision is served. This is why the trend of mediations in the context of boundary disputes emerged with the aim that the parties enter into a deal, with the idea that there are no winning or losing parties. This is elaborated on in Chapter IV. However, there are situations where parties simply cannot come to an agreement. In such cases, usually the county court would be the most common forum for such disputes.[279]

However, the proposed process relies entirely on the expertise of surveyors. Therefore affected parties should also seek legal advice in relation to issues involved in boundary disputes including, for example, legal document interpretation, the concept of estoppel, boundary agreements and the issue of adverse possession.[280] The same applies to the PWA 1996 considering that, again, relevant surveyors are not regulated.

In addition, the Bill, if passed, would not apply to boundary disputes between leasehold owners of land or between a leasehold owner and a freehold owner of land (unless the freehold owner was willing to be involved in the procedure imposed by the Bill).

This is because the Bill defines the 'owner of land' as a:

> ... *freehold owner of land who is desirous of establishing the position of a boundary between his land and the land of an adjoining owner or a private right of way.*[281]

---

278 C Sara, D Dovar, *Boundaries and Easements*, 7th edn (Sweet & Maxwell, 2019) 24-001.

279 A Moor, 'Property Litigation Blog; Boundary disputes' (2017) *Practical Law UK Articles*, Resource ID W-008-6296, https://uk.practicallaw.thomsonreuters.com/Document/I57332ceb51de11e798dc8b09b4f043e0/View/FullText.html?transitionType=SearchItem&contextData=(sc.Search).

280 ibid.

281 Property Boundaries (Resolution of Disputes) Bill [HL], s 15.

An 'adjoining owner' refers, like the 'owner', to a freehold owner of land. Boundary disputes between freehold owners can still end up in court through the appellate procedure, where a party is not happy with a surveyor's decision.

The Bill is unlikely to move forward. Boundary disputes, like party wall disputes, suffer from similar problems. For example, it is not uncommon for the value in dispute to be lower than or similar to the costs linked to court proceedings. Again, if surveyors are chosen to decide such disputes, the issue is that they are not regulated, which means that there is a variability in quality and direction of their awards. While this is far from ideal, the trend for the future is to prevent courts being unnecessarily burdened with boundary and party wall disputes, and for these to be solved (if not through the PWA 1996 dispute resolution mechanism in relation to party walls) with ADR, leading to parties having more control over the process and outcome as well as linked cost and time in relation to the relevant dispute.

5.5 Human Rights Act 1998

The Human Rights Act 1998 (**HRA 1998**) enshrines two main rights that can affect party walls and related disputes: right to a fair and public hearing within a reasonable time by an independent and impartial tribunal established by law (Article 6(1)) and respect for one's private and family life, one's home and one's correspondence (Article 8). These rights apply also to party walls and related disputes and are therefore analysed further below.

The purpose of the HRA 1998 is that it incorporates most of the provisions of the European Convention for the Protection of Human Rights and Fundamental Freedoms agreed by the Council of Europe at Rome on 4 November 1950 (**Convention**) into domestic law. The most relevant articles of the Convention to party walls are listed below as well as the reasons as to why they are relevant.

5.5.1 Article 6(1)

Article 6(1) of the Convention states:

> *In the determination of his civil rights and obligations or of any criminal charge against him, everyone is entitled to a fair and public hearing within a reasonable time by an independent and impartial tribunal established by law.*

Article 6(1) of the Convention is likely to apply to a party wall award as this can be viewed as a determination of civil rights and obligations of the parties to the dispute. This is because of the consequences the award will have to which the PWA 1996 applies. The main issue here is whether the system set out in the PWA 1996 for determining party wall awards complies with Article 6(1) of the Convention. A problem could arise if there were limitations as to the recourse to a formal judicial determination.[282]

---

282 Bickford-Smith, Nicholls and Smith (n 22) 15.33.

This issue was addressed in the case of *Zissis v Lukomski*,[283] where the claimant and the first defendant were owners of adjoining houses. The claimant wanted to carry out works covered by the PWA 1996 and therefore issued a notice to the first defendant under section 6(5) of the PWA 1996. The case revolves around an appeal to the county court against a party wall award under section 10(17) of the PWA 1996, which was a statutory appeal governed by CPR Part 52.[284] It was held that the appeal court had ample powers under Part 52 so that it can determine the appeal in a just way.[285]

The dispute related to the works. The second defendant, acting as the first defendant's surveyor pursuant to section 10(1)(b) of the PWA 1996, made an addendum award under section 10(7) of the PWA 1996. The claimant then appealed against the award pursuant to section 10(17) of the PWA 1996 by way of commencing proceedings under CPR Part 8. The claimant sought orders that the award be rescinded or modified. The district judge dismissed the claim, despite asserting that the addendum award was not valid. The judge held that the claimant's appeal was a statutory appeal and therefore should have been brought under CPR Part 52. As the claimant however was out of time in relation to an appeal under CPR Part 52, the judge dismissed the claim.[286] The claimant appealed to the Court of Appeal, which held that the claimant appealing the award in the county court was not correct procedurally as appeals under section 10(17) of the PWA 1996 need to be made under CPR Part 52 and not under Part 8.

> *I think it plain that CPR Pt 52 was intended to cover a form of statutory appeal like that under section 10(17) and that the provisions of CPR Pt 52 are amply sufficient to allow justice to be done on such an appeal.*[287]

As the award was made under the PWA 1996 without a hearing, an appeal by way of a rehearing would mean that it was open to the county court to accept fresh evidence before reaching its own independent decision as to whether the award was incorrect. CPR Part 52 was held to be appropriate to ensure justice is served on an appeal under section 10(17) of the PWA 1996. The Court of Appeal also held that the district judge was incorrect in dismissing the claim considering that he concluded the addendum award was invalid. Instead, the district judge should have allowed for the claimant's particulars of claim to be amended so that they sought relief in the form of a declaration that the award was a nullity. Alternatively, the district judge should have allowed for the proceedings to continue under CPR Part 52(3).[288]

> *Submissions to the effect that an appeal hearing should be a rehearing are often motivated by the belief that only thus can sufficient reconsideration be given to elements of the decision of the lower court. In my judgment, this is largely unnecessary given the*

---

283 (n 128).

284 'Party Wall etc Act 1996 – appeal procedure under section 10(17)' (2006) 22(7) *Construction Law Journal* 469–481.

285 *Zissis v Lukomski* (n 128).

286 ibid. 18.

287 ibid. 40.

288 ibid.

scope of a hearing by way of review under rule 52.11(1). Further the power to admit fresh evidence in rule 52.11(2) applies equally to a review or rehearing. The scope of an appeal by way of a review, such as I have described, in my view means that the scope of a rehearing under rule 52.11(1)(b) will normally approximate to that of a rehearing 'in the fullest sense of the word' such as Brooke LJ referred to in para 31 of his judgment in Tanfern Ltd v Cameron-MacDonald (Practice Note) [2000] 1 WLR 1311. On such a rehearing the court will hear the case again.... Circumstances in which the hearing of an appeal will be a rehearing are described in paragraph 9 of the Practice Direction supplementing Part 52. This refers to some statutory appeals where the decision appealed from is that of a person who did not hold a hearing or where the procedure did not provide for the consideration of evidence. In some such instances, it might be argued that the appeal would in effect be the first hearing by a judicial process, and that a full hearing was necessary to comply with article 6 of the European Convention on Human Rights and Fundamental Freedoms ... It will if necessary hear evidence again and may well admit fresh evidence.[289]

The Court of Appeal further held that PWA 1996 awards cannot be enforced by way of summary procedure set out in CPR 70.5.

CPR r 70.5 creates a procedure, which can be initiated by a court officer without the intervention of a judge, for the enforcement of an award of a sum of money made by any court, tribunal, body or person other than the High Court or a County Court so long as an enactment provides that the award may be enforced as if payable under a court order, or that the decision may be enforced as if it were a court order. The 1996 Act contains no such provision, so that an award made under that 1996 Act cannot be enforced through the CPR r 70.5 procedure.[290]

As either party can appeal against the award, through which it has access to a full rehearing in the county court (section 10(17) of the PWA 1996), it is unlikely that the system will be criticised for not complying with Article 6(1) of the Convention.[291] Article 6(1) of the Convention and its effect as to whether a statutory provision that has severe consequences in a specific case should be declared incompatible with the Convention right was discussed in the House of Lords decision in *Wilson v First County Trust Ltd.*[292] It was held that the harsh consequences caused by a statutory provision will not result in the provision being declared incompatible with the relevant Convention right.[293]

Section 6(1) provides that it "is unlawful for a public authority to act in a way which is incompatible with a Convention right." On a natural reading this provision is directed at post-Act conduct. The context powerfully supports this interpretation. One would not

289 ibid. 39.
290 ibid 59.
291 Bickford-Smith, Nicholls and Smith (n 22) 15.33.
292 [2004] 1 A.C. 816.
293 Bickford-Smith, Nicholls and Smith (n 22) 15.38.

> *expect a statute promoting human rights values to render unlawful acts which were lawful when done. That would be to impose liability where none existed at the time the act was done.*[294]

The reference in the above quotation is to section 6(1) of the HRA 1998. The argument here is that where party wall awards have been made in contravention of this section, as long as the award preceded the HRA 1998, the HRA 1998 will not apply. The emphasis is therefore on awards under the PWA 1996 once the HRA 1998 came into force on 2 October 2000 incorporating the Convention provisions in England, Wales and Northern Ireland.

Before the PWA 1996, under the London Building Acts (Amendment) Act 1939, Brightman J. noted in *Gyle-Thompson and Others v Wall Street (Properties) Ltd.*:[295]

> *The position of the adjoining owner whose proprietary rights are being compulsorily affected, is intended to be safe- guarded by the surveyors appointed pursuant to the procedure laid down by the Act. Those surveyors are in a quasi-judicial position with statutory powers and responsibilities. It therefore seems to me important that the steps laid down by the Act should be scrupulously followed throughout and short cuts are not desirable.*

The Earl of Lytton described the role of the surveyor to be quasi-arbitral at the time he introduced the Bill to the House of Lords:[296]

> *The duty of party wall surveyors is quasi-arbitral. Once appointed they have a duty to act properly in the interests of both parties as statutory surveyors, which is a most important safeguard.*

In *Mills v Savage*, HHJ Bailey noted:[297]

> *Party wall surveyors are exercising a quasi-arbitral function. They are bound by the rules of natural justice. It is axiomatic that in considering and making an award a party wall surveyor, and this must include the third surveyor, must enable the parties to make submissions if they wish and must give due consideration to any submissions made.*

Indeed, the surveyor must act independently, meaning that the appointing owner cannot control the surveyor by, for example, threatening them with loss of their appointment. The party wall surveyor must also act impartially and professionally.[298] In allowing each owner

294 *Wilson v First County Trust Ltd* (n 292) 12.
295 (n 131) 130.
296 Lord Lytton introducing the Party Wall, Etc Bill on its second reading, which led to the PWA 1996 in the House of Lords Debate on 31 January 1996 (n 57).
297 (n 117), 131.
298 *Welter v McKeeve* (n 113), 165.

to appoint a party wall surveyor, the PWA 1996 is not inviting party-appointed surveyors to act partially without regard to the merits of the material before them, and simply pursue the interests of the appointing party. Rather, it is helping to ensure that the proper concerns and interests of both owners are raised for due and impartial consideration in the dispute resolution process. In other words, the party wall surveyor is 'acting for the wall'.

The analysis of the relevant legislation provision has to be done in context, and it has to be assessed whether the set of measures is a proportionate response as to the issue at hand by Parliament.[299] While Section 6(1) of the HRA 1998 is unlikely to cause an issue, when going through the dispute resolution motions, parties would be wise to consider from a strategic perspective whether, should one of them opt for an appeal, rehearing is something that could benefit their case. A rehearing has the advantage of, for example, further evidence being considered by the judge, which can affect or even result in a change of the outcome of the judgment.

5.5.2 Article 8

According to Article 8 of the Convention:

> *(1) Everyone has the right to respect for his private and family life, his home and his correspondence.*
> *(2) There shall be no interference by a public authority with the exercise of this right except such as is in accordance with the law and is necessary in a democratic society in the interests of national security, public safety or the economic well-being of the country, for the prevention of disorder or crime, for the protection of health or morals, or for the protection of the rights and freedoms of others.*

It is clear that activities under sections 2 and 6 of the PWA 1996 can lead to interfering with home and family life. This is because works related to party walls can cause noise, or vibration or create dust, which can sometimes be disruptive. Pollution and noise may lead to infringement of Article 8 of the Convention.[300] However, whether this actually amounts to such an infringement depends on the level of the disruption. If losses of amenity are minor, this is unlikely to lead to interference with Article 8 of the Convention.[301] Those affected by such interference would be the adjoining owner and Adjoining Occupier. This is because these parties have no choice but to submit to the building owner's activities. This is subject to an award under section 10 of the PWA 1996 (for the adjoining owner).[302]

Relevant provisions of the PWA 1996 do not interfere with Article 8 of the Convention; however, the factors set out below need to be taken into consideration:[303]

> *(1) The circumstances in which interference can occur are in practice limited.*
> *(2) The rights of the building owner are closely defined and circumscribed.*

---

299 Bickford-Smith, Nicholls and Smith (n 22) 15.33.
300 *Lopez Ostra v Spain (A/303-C)* (1995) 20 E.H.R.R. 277; *Hatton v United Kingdom (36022/97)* (2003) 37 E.H.R.R. 28.
301 *Lough v First Secretary of State* [2004] 1 WLR 2557.
302 Bickford-Smith, Nicholls and Smith (n 22) 15.34.
303 ibid. 15.36.

(3) Compensation is payable on exercise of those rights.
(4) In the case of an adjoining owner, he must be notified of the works proposed and is entitled to an award from an independent surveyor or tribunal of surveyors.
(5) The interference is transitory, lasting only as long as the works in question.
(6) The statutory scheme as a whole may be justified on the basis of economic well-being, and is proportionate.

As to whether there could be a situation where particular exercise of the powers under the PWA 1996 could lead to a breach of Article 8 of the Convention, it is not likely that the exercise of the rights under section 2 or 6 of the PWA 1996 in relation to the position of an adjoining owner would lead to a claim that Article 8 of the Convention was engaged. However, this does not apply to Adjoining Occupiers as they are not entitled to take part in the procedure. It follows that it is the responsibility of building owners that Adjoining Occupiers are not affected unreasonably by noise, pollution or discomfort caused by the building owner's exercise of its rights under the PWA 1996. As for rights of entry under section 8 of the PWA 1996, these are not likely to lead to a breach of Article 8 of the Convention as these provisions (such as a warrant) facilitate a speedy resolution of party wall disputes and they are not required to be exercised in any specific way, or at all.[304]

On balance, section 8 of the PWA 1996 relating to rights of entry should not be exercised too harshly as such conduct risks later being held in breach of Article 8 of the Convention by court. This is because the consequences of such conduct would not immediately give rise to a new cause of action under the HRA 1998, which only requires public authorities to protect human rights. However, as such rights of entry under section 8 of the PWA 1996 are restricted to 'necessity', the probability of these being exercised too harshly is unlikely. A party, however, that feels its rights have been or are likely to be violated, can apply to the county court for relief against the party in breach. Theoretically, a party can start proceedings in court seeking, for example, an injunction or appeal against a party wall award during which it can raise an Article 8 of the Convention argument, which can impact on the relief granted by the court.[305]

5.5.3 Article 1 of the First Protocol to the Convention

Article 1 of the First Protocol to the Convention (**First Protocol**) states:

> *Every natural or legal person is entitled to the peaceful enjoyment of his possessions. No one shall be deprived of his possessions except in the public interest and subject to the conditions provided for by law and by the general principles of international law.*
>
> *The preceding provisions shall not, however, in any way impair the right of a State to enforce such laws as it deems necessary to control the use of property in accordance with the general interest or to secure the payment of taxes or other contributions or penalties.*

304 ibid. 15.34-15.39.
305 ibid. 15.36-15.39.

Although works performed under the PWA 1996 can lead to the interference with enjoyment of a party's possessions, the PWA 1996 is almost certainly compliant with Article 1 of the First Protocol due to the safety measures contained in the PWA 1996,[306] such as the time limitation on the duration of an interference as well as compensation provisions.[307] The PWA 1996 also has public benefit as its core aim.[308] Additionally, one has to consider that by section 1(6) of the PWA 1996:

> *Where the building owner builds a wall wholly on his own land in accordance with subsection (4) or (5) he shall have the right, at any time in the period which–*
>
> *(a) begins one month after the day on which the notice mentioned in the subsection concerned was served and*
>
> *(b) ends twelve months after that day,*
>
> *to place below the level of the land of the adjoining owner such projecting footings and foundations as are necessary for the construction of the wall.*

This could be interpreted in such a way that the PWA 1996 essentially allows for subterranean encroachment. This could amount to an infringement of Article 1 as it could be viewed that the building owner is expropriating the adjoining owner's rights in its subterranean property. One could argue that the main issues are whether the encroachment is necessary for the construction of the wall as well as the degree of such encroachment.[309] However, these factors are opaque, and it remains to be seen what courts will decide should such issues be raised.

5.6 Environmental Protection Act 1990 and Control of Pollution Act 1974

According to DCLG's Explanatory Booklet:[310]

> *You should contact your local authority environmental department who have powers under the Environmental Protection Act 1990 and the Control of Pollution Act 1974 to deal with matters of noise and other potential nuisance, such as dust and deposits from construction sites.*

The Environmental Protection Act 1990 (**EPA 1990**) deals with noise nuisance and its main aim is to ensure that the environment quality is not reduced via excessive noise,

306 ibid. 15.40.
307 *Stran Greek Refineries v Greece (A/301-B)* (1995) 19 E.H.R.R. 293; *Matos E Silva Lda v Portugal* (1997) 24 E.H.R.R. 573.
308 *James v United Kingdom (A/98)* (1986) 8 E.H.R.R. 123.
309 P Aliker, 'Party walls and arbitration' (2005) *New Law Journal* 155 NLJ 866.
310 (n 26) 40.

meaning noise that is loud and intrusive, whether persistent or intermittent, affecting the quality and comfort of everyday life (such as construction works to party walls). The EPA 1990 also deals with other statutory nuisances relevant to party walls, such as:

a) smoke, fumes or gases from any premises;
b) dust, steam or smells from business premises;
c) accumulations or deposits;
d) noise or vibration;
e) premises in a poor state; or
f) artificial light.

For these to be considered statutory nuisances, they have to pose a health risk or a nuisance. Where a party is not able to address the nuisance issue directly, it has the option of contacting its local authority Environmental Health Department or a similar authority, who then has to investigate the complaint and then take action on the complainant's behalf if the authority comes to the conclusion that a statutory nuisance has materialised, is likely to materialise or be repeated.[311]

Theoretically, where the relevant authority does not feel it can take action, the affected party can take action through the magistrate's court under section 82 of the EPA 1990, which is a swift and a relatively cheap way of dealing with the issue as no legal representation is necessary.

The Control of Pollution Act 1974 (**COPA 1974**) relates to control of noise on construction sites, which is determined by the relevant local authority. The authority will serve a notice stipulating how noise is to be controlled where advance consent has not been secured. Conditions are usually imposed as part of a planning permission but in a very generic form. Invariably the controls imposed on notifiable works within party wall awards tend to be both more comprehensive and stricter.

The COPA 1974 provides for a definition of best practicable means to minimise noise and vibration. Sections 60 and 61 of the COPA 1974 ensure that all construction sites are subject to a notice and consent procedure where a party can apply for consent in advance from the neighbouring party to be able to perform activities generating noise during construction works.[312]

Examples of works to party walls that may be caught by the COPA 1974 include: "... *use of power tools that utilise percussive, boring, cutting, grinding or impact techniques on party walls/floors or occupied properties ...* "[313]

---

311 Environmental Protection Act 1990, s 79.
312 Royal Borough of Kensington and Chelsea, 'Code of construction practice: Minimising the impact of noise, vibration and dust' (2016), 43, https://www.rbkc.gov.uk/environment/advice-builders/code-construction-practice-minimising-impact-noise-vibration-and-dust.
313 ibid. 21.

5.7 Crossrail Act 2008, High Speed Rail (London – West Midlands) Act 2017 and High Speed Rail (West Midlands – Crewe) Act 2021

5.7.1 Crossrail Act 2008

Following the implications of the Channel Tunnel Rail Link Act 1996 (which modified similar legislation, which the PWA 1996 replaced), the Crossrail Act 2008 has legislative powers to disapply the effect of the PWA 1996. According to section 17 of the Crossrail Act 2008:

> *(1) No notice under section 1(2) or (5) of the Party Wall etc. Act 1996 (c. 40) (notice before building on line of junction with adjoining land) shall be required before the building of any wall in exercise of the powers conferred by this Act.*
> *(2) Sections 1(6) and 2 of the Party Wall etc. Act 1996 (rights of adjoining owners) shall not have effect to confer rights in relation to–*
> *(a) anything held by the Secretary of State or the nominated undertaker and used, or intended for use, by the nominated undertaker for the purposes of its undertaking under this Act, or*
> *(b) land on which there is any such thing.*
> *(3) Section 6 of the Party Wall etc. Act 1996 (underpinning of adjoining buildings) shall not apply in relation to a proposal to excavate, or excavate for and erect anything, in exercise of the powers conferred by this Act.*

However, the Crossrail Act 2008 provides for mitigating measures with the aim to protect structures located in areas that are at a high risk of ground settlement.[314]

5.7.2 High Speed Rail (London – West Midlands) Act 2017 and High Speed Rail (West Midlands – Crewe) Act 2021 (HS2 Legislation)[315]

The purpose of the amendments to the PWA 1996 by the HS2 Legislation is to ensure that there are safeguards in place for owners of buildings or land adjacent to the HS2 railway. At the same time, the aim is also to ensure a safe cost- and time-effective delivery and operation of the HS2 railway. Excavation as well as underpinning works and party wall works are covered.[316] The changes were introduced as otherwise, each interface of the HS2 works with an adjacent property would need to be determined in line with the procedures in the PWA 1996 or agreed between surveyors. Another purpose of the changes was to reduce the likelihood of complex disputes with adjoining owners while also protecting them. Schedule 23 of the HS2 Legislation specifically provides amendments to the notification procedure

314 S Waldron, 'Crossrail: lifting legislative provisions – Easing the load' (RICS isurv, 2017), https://www.isurv.com/info/390/features/11162/crossrail_lifting_legislative_provisions.
315 HS2, 'Party wall guidance under the HS2 regime' (2022), www.hs2.org.uk.
316 ibid. 1.

(sections 1(6) and 2 of the PWA 1996), rights of adjoining owners (section 3 of the PWA 1996), underpinning of adjoining buildings (section 6 of the PWA 1996) and dispute resolution procedure, where a modified arbitration procedure applies if a dispute comes about in relation to an issue connected with works to which the PWA 1996 would otherwise have applied and the work is needed for HS2 purposes or relates to buildings/structures located on Secretary of State's or the Nominated Undertaker's land in order to build the HS2 railway.[317]

5.8 Conclusion

When considering legal coherence and dispute management in relation to party-wall-related disputes, party walls must be seen in a legislative context. The PWA 1996 is not the only piece of legislation that can apply to party walls, which is the reason why the ANLA 1992, the LPA 1925, the Bill, the HRA 1998 and the Convention, as well as the EPA 1990, the COPA 1974, the Crossrail Act 2008 and the HS2 Legislation are discussed in this book.

The PWA 1996 does not authorise infringement of other pieces of legislation. However, both the Crossrail Act 2008 and the HS2 Legislation provide examples of where an adjoining owner's usual rights under the PWA 1996 are overridden or varied. Where parties are in a party wall dispute, they must not rely automatically on the PWA 1996 alone. In order to prevent a dispute or manage it, if it cannot be avoided, building owners must be aware what legislation they need to comply with when performing works under the PWA 1996.

The ANLA 1992 should be viewed as potentially complementary with the PWA 1996, linking into practical issues of scaffolding and crane oversail to ensure that the building owner does not commit trespass.

Boundary determination is intrinsically linked with party walls, which can stand on or be linked to boundary lines and, in this context, the LRR 2003 and LRA 2002 need to be borne in mind. In the context of boundary determination, the LPA 1925 needs to be considered as it is relevant both to boundaries and party walls.

The Bill, while its future is uncertain, is a step in a similar direction as the PWA 1996 in that it stems from the same mind-set and aims to assist neighbouring parties to resolve disputes without the need to go to court straightaway.

The HRA 1998 and Convention can affect relationships between neighbouring parties when it comes to party walls and related disputes, specifically Articles 6 (entitlement to a fair and public hearing) and 8 (right to respect for one's private and family life) of the Convention and Article 1 of the First Protocol to the Convention (peaceful enjoyment of one's possessions). The HRA 1998 incorporates directly the most important sections of the Convention into British domestic law. It affects interpretation and use of all legislation in Britain and therefore it affects interpretation and use of the PWA 1996.

Article 6(1) of the Convention, which directly refers to the right to fair and public hearing within a reasonable time by an independent and impartial tribunal, was further examined in *Zissis v Lukomski*.[318] As a consequence, it is clear that either party is able to appeal against a party wall award and so has recourse to a full rehearing in the county court

317 ibid. 3.
318 (n 128).

according to section 10(17) of the PWA 1996. As a result, the likelihood that the system under the PWA 1996 will be criticised for non-compliance with Article 6(1) of the Convention is small.[319]

In connection with Article 6(1) of the Convention, it was also considered whether a statutory provision that has severe consequences in a specific case should be declared incompatible with the Convention right in *Wilson v First County Trust Ltd.*[320] The conclusion is that where party wall awards have been made in contravention of this section, as long as such awards preceded the HRA 1998, the HRA 1998 will not apply. The emphasis is therefore on awards under the PWA 1996 once the HRA 1998 came into force on 2 October 2000 incorporating the Convention provisions in England, Wales and Northern Ireland.

Article 8(1) of the Convention relates to the right to respect for private and family life, home and correspondence. Activities under sections 2 and 6 of the PWA 1996 can interfere with Article 8(1) of the Convention if levels of pollution and noise are excessive. A party can start proceedings during which it can raise an Article 8 of the Convention argument.

Although the relevant sections (2, 6 and 8) of the PWA 1996 do not appear to clash with Article 8 of the Convention, there are a number of factors that need to be taken into consideration. While exercise of the rights under section 2 or 6 of the PWA 1996 in relation to the position of an adjoining owner is not likely to give rise to a claim that the Article 8 of the Convention was engaged, it is important to point out that this does not apply to Adjoining Occupiers. Therefore, it is the building owners' responsibility that Adjoining Occupiers are not affected unreasonably by noise, pollution or discomfort caused by the building owners' activities under the PWA 1996. In addition, rights of entry under section 8 of the PWA 1996 should also not be exercised too harshly to ensure that Article 8 of the Convention is complied with.

Works performed under the PWA 1996 are almost certainly compliant with Article 1 of the First Protocol as the PWA 1996 contains safety measures mitigating circumstances that could lead to infringement of Article 1 of the First Protocol. In addition, the PWA 1996 has public benefit as its core aim. A building owner should however be cautious when performing works under section 1(6) of the PWA 1996 as these could amount to expropriation of the adjoining owner's rights in its subterranean property (where relevant).

Finally, in the context of potential nuisance caused by works performed on a party wall, the EPA 1990 and COPA 1974 need to be borne in mind. The Crossrail Act 2008 and the HS2 Legislation have an impact on the PWA 1996 in that they amend its provisions and/or application in the context of Crossrail and HS2.

The message for building and adjoining owners is therefore to think contextually, not to get comfortable under the seemingly 'protective shield and guidance' of the PWA 1996 and not to be lured into a false sense of security. It is crucial for the parties to be aware of their rights and obligations in order to prevent or effectively manage potential disputes.

319 Bickford-Smith, Nicholls and Smith (n 22) 15.33.
320 (n 292).

6

Proprietary Rights in Land – Easements

6.1 Introduction – Easements

When discussing the definition of 'party wall' in Chapter II, reference was made to four categories used prior to 1926 as to what the term 'party wall' may mean. Two of those categories related to easements, i.e.:[321]

> *Before 1926 the term "party-wall" (in relation to a wall separating two properties) might mean: (1) a wall of which the two adjoining owners were tenants in common; or (2) a wall divided longitudinally into two portions, one belonging to each adjoining owner; or (3) a wall belonging entirely to one adjoining owner, but subject to an easement in the other to have it maintained as a dividing wall; or (4) a wall divided longitudinally into two portions, each portion being subject to a cross-easement in favour of the owner of the other.*

While the effect of the LPA 1925 has been to transfer walls, which prior to the PWA 1996 would have been in the first of the four categories, into the fourth category, party walls have been historically linked to easements, as they presently also are.[322]

While party walls are not easements themselves, easements can affect party walls. This is why this chapter is dedicated to several types of easements in the context of party walls. The main easements particularly relevant to party walls are: a) rights of way; b) rights of support; c) drainage rights; and d) rights to light (however this is covered in Chapter VII in relation to obstruction of passage of light leading to a nuisance in tort). While the right or obligation to repair is not an easement as such, it is relevant to easements and can be intrinsically linked to these as discussed further in this chapter.

321 *Gale on the Law of Easements* (n 11) 11-01.
322 ibid. 11-02.

Party Wall Disputes: The Party Wall etc. Act 1996 and Beyond – *Legal Coherence and Dispute Management*, First Edition. Dr Laura Lintott.

6.2 Easements – Right of Way

Considering that party walls can suffer deterioration leading to collapse, they need to be considered in the context of the easement of right of way (for example, where a footpath runs alongside a party wall and there is a public right of way). Liability could arise for the party wall owner(s) where, for example, the party wall falls and causes personal injury to a by-passer. Where there is a grant of an easement over a part of land, this will usually not entitle the user to demolish parts of the party wall for the purpose of widening the access unless express provisions allow for this.[323]

The examples of case law set out below show the interface that can arise between the easement of a right of way and party walls. They also show examples simply where a party wall co-exists in close proximity to a right of way.

In the case of *Mills v Blackwell*,[324] there was a field whose owner sold the roadway, however reserved a right of way at all times and for all purposes connected with the present and every future use of the field. This included vehicles as well. The roadway and the field were separated from each other by a stone wall, which contained a wooden gate. Both properties changed ownership, and the rights and burdens were passed to the respective successors in title. The now new owner of the field decided to demolish part of the stone wall for the purposes of adding six feet to the width of the gateway. The owner of the right of way objected and claimed that there was no entitlement to demolish sections of what was essentially a party wall and that there was also no entitlement to broaden the gateway.

It was held that the "*conveyance had to be construed in accordance with the physical characteristics of land at the time.*"[325] The reason behind this conclusion was the fact that there was no specific provision in relation to the way in which the gateway was to be treated apart from the declaration that the wall was a party wall.

The Court of Appeal held that there was no entitlement to demolish any part of the party wall or to broaden the gateway. Therefore, the previous position had to be reinstated. The issue here was that the reservation was 'badly worded' and failed to deal with the point at which the right of way entered the adjoining land. If the wording of the reservation in the original conveyance covered the possibility, the position would have been clear and the problem could have been mitigated in this way.

> *The conveyance, moreover, did not specifically mention any points relating to access or egress. Consequently, it would be absurd to conclude that Mr Blackwell was entitled to demolish the whole of the wall so that access and egress might be obtained from any point along the strip "But why should they be entitled to choose an access point anywhere they may reasonably select, when it is absolutely plain from the physical layout at the time of the conveyance that the access point was at and through the gate 4 feet 6 inches wide." There was nothing insubstantial or transient about a dry stone wall and*

323 Case Comment, 'Easements: point of access' (1999) 20(3) *Property Law Bulletin* 18–19.
324 (1999) 78 P. & C.R. D43.
325 ibid.

> *simply nothing in the conveyance to suggest an intention that the point or extent of access or egress should be anywhere or be to a greater extent than what was then capable of enjoyment*[326]

In *Dawoodi v Zafrani*,[327] the claimant owned a property that included a narrow path. The original purpose of the path was to allow deliveries to the back of the adjacent properties. The path was however no longer used for the purpose for which it was built. As the path was no longer in use when the claimant purchased the property, she decided to build a large shed extending onto the path. The first defendant bought the neighbouring property with the plan to demolish the property he bought and to build a new building in its place. His architect provided incorrect advice to him that the claimant's path was within the first defendant's boundary. Following this advice, the first defendant started works on his property without having applied for planning permission, without serving a party wall notice and without the consent of the local authority's building control. As part of the works, he also built scaffolding on the claimant's path, again, without asking the claimant for permission to do so.

The court held:

> *Of course, the defendant should have obtained a party wall award. In such an award, a licence could have been granted for the use of the claimant's land. A fee would have been payable for that use.*[328]

It has to be noted however that if a party wall award had been obtained authorising the relevant work, no licence would have been necessarily required due to rights of entry/access under section 8 of the PWA 1996. A party wall can interfere with the right of way and, in order to prevent a dispute between the parties, the best way forward is proper communication, which in this case would have included the defendant issuing a party wall notice to the claimant. In this way, court proceedings could have been avoided if the defendant had taken steps to keep the claimant informed of what it intended to do, giving the claimant the opportunity to agree or disagree in advance of any works and finding a solution that would be acceptable to both parties in an amicable way contained in a party wall agreement or, if not acceptable for the parties, by way of a party wall award.

The case of *Parkinson v Reid*[329] revolves around a right of way that refers to a stairway. In 1926, the owner of lot 28 granted the owner of lot 29 a right of way over the abovementioned stairway. The stairway led to the second floor of the building on lot 28. The owner of lot 28 covenanted with the owner of lot 29 to keep the stairway in repair and reconstruct it should the stairway be partially or completely destroyed. At the same time, the owner of lot 29 granted the owner of lot 28 the right to use the flank wall on lot 29 as a party wall. The point was that the party wall would support the stairway. The titles of both lots passed

---

326 ibid. D44.
327 [2015] EWHC 3168 (TCC).
328 ibid. 58.
329 [1966] 1 WLUK 544.

on to different parties and in 1961, lot 28 was affected by fire in which the building was badly damaged. As a result, the damaged building and the stairway, which was not itself damaged, were pulled down. The claimant however decided to seek a mandatory injunction to ensure that the covenant to repair the stairway would be enforced. The court held that although there was a covenant to repair and reconstruct the stairway, its burden did not pass on with the land and therefore was not enforceable as privity of contract was missing. The benefit of the covenant to enjoy the use of the wall on lot 29 as a party wall passed on with the land. As long as the stairway was used, the owner of lot 29 had the duty to keep the stairway in repair. However, the obligation ended when the use ended. As the staircase was removed, its use ceased and therefore the court concluded that the court action had to fail.[330]

In this case, the party wall was supporting the right of way (staircase). The party wall did not negatively interfere with the right of way; however, it formed part of it. It is another point of view where the interplay between a party wall and a right of way is in itself not contentious. This case focussed more on the fact that the obligation to maintain the right of way ceased with the end of use of the staircase. However, the staircase could not have existed on its own without the party wall, which suggests that a relationship between a party wall and a right of way is possible.

In the case of *Burlington Property Co Ltd v Odeon Theatres Ltd,*[331] which is a Court of Appeal case, the defendant owned 33 Charing Cross Road, and the claimant owned the adjoining premises at 35 Charing Cross Road. The flank wall of the defendant was a party wall, which was several storeys high. On the other side of the party wall was the claimant's land (Hunt's Court), which was subject to a public right of way. The defendant planned to demolish the party wall and rebuild it with arches allowing access to no. 33 from Hunt's Court.[332] Surveyors were appointed as 'arbitrators'[333] to settle the differences between the claimant and the defendant. The surveyors' award stated that the defendant was entitled to go ahead with its proposed works. At first instance, the court held that the wall was in fact a party wall and re-affirmed the validity of the surveyors' award. The claimant appealed. The Court of Appeal (Slesser L.J.) stated:[334]

> *... it appears by the conveyance of 1905 that at that date there was a party wall and that the half of the wall, with regard to which complaint is now made that it has been interfered with, was and has been at all material times the property of the appellants. In those circumstances, unless the respondents here can show some statutory right to interfere with what, on the face of it, is the absolute property of the appellants, it must necessarily be that an award which gives to them a power to do that which in law they have no power to do must be bad. I seek in vain in the only statute suggested to be*

330 ibid.
331 (n 275).
332 London Building Act 1930, section 117(1).
333 ibid. Part IX.
334 *Burlington Property Company, Limited v Odeon Theatres, Limited* (n 275), 641.

> *competent to give them the right for any power whatever in this behalf. Sub-sect. (vii.) of s. 114 of the London Building Act, 1930, seems to me to give them no such right. It confers a right "to pull down any party structure which is of insufficient strength for any building intended to be built and to rebuild the same of sufficient strength for the above purpose." The whole complaint which the appellants make is that their moiety of the party structure has not been rebuilt as it formerly was, and certainly has not been rebuilt with any consideration of strength, seeing that the apertures in it are considerably wider than they were in the original building.*

The Court of Appeal held that the surveyors lacked jurisdiction to make such an award and that the defendant could only demolish and rebuild the party wall in accordance with the substantially same design as it was. Therefore, the defendant was not entitled to rebuild the party wall to a different design.

This is a case showing that there are situations where a party wall can run along a right of way (in this case a public right of way). While a right of way and party wall can exist independently from each other in close proximity, they can interact or be relevant to each other where, for example, the right of way crosses a party wall or where a party wall falls into disrepair affecting the right of way.

While the dispute was about whether one party was able to demolish and rebuild the party wall based on different designs compared to its original designs, the fact remains that the crux of the issue was ensuring that the party wall was of the same strength to support the respective buildings. This case highlights the status of a 'party wall' as opposed to 'any' wall in that interference with a party wall must follow a given set of rules, either the parties can agree between themselves how to deal with the party wall or alternatively, a party wall award is necessary to ensure works to a party wall are carried out appropriately. While on the face of it the defendant was acting in accordance with the party wall award, the issue here was that the surveyors went beyond their jurisdiction in the first place. It is an interesting case in that the building owner was following an invalid party wall award.

In conclusion, party walls can interfere with rights of way. They can also be interdependent. A party wall can form a part of a right of way. It can demarcate the boundary of the easement. When a party wishes to build, demolish or change a party wall, it should carefully investigate its rights and obligations, whether the wall is a party wall and communicate with the affected neighbouring party(/ies) before starting works, to try and prevent a potential dispute. It is advisable for the parties to keep a record of their negotiations and any settlement attempts, which could be considered as a mitigating factor by courts should the situation escalate. Any agreement between the parties as to how to proceed with works on a party wall should be recorded in writing for future reference and potential use in court to prevent any inconsistencies or miscommunications and establish certainty of what the parties agreed. Where parties cannot agree how to deal with a party wall, they should seek a party wall award deciding the matter. From a practical perspective, the building owner should think carefully before carrying out any works to the party wall where the adjoining owner is appealing the party wall award until the matter is fully resolved. This is to mitigate any loss of money and time.

6.3 Easements – Right of Support

6.3.1 Introduction

When discussing the right of support, a distinction needs to be made between the right to the support to structures built on the land and a natural right regarding support to land in its natural state, which exists automatically (it arises naturally for land) and enjoys the protection of the law of tort. A natural right of support cannot be removed and if the support is removed leading to subsidence, the affected party can seek damages.[335] A natural right of support is enjoyed by every landowner who may have an action against a neighbour in nuisance for deliberately removing the support by quarrying[336] or by having removed a subjacent strata of minerals[337] or where the support is removed by natural causes arising on the neighbour's land (where the defendant was aware of the danger and its failure was in the fact that it did not act reasonably and did not remove the hazard).[338] It is only the land in its natural state to which the natural right of support extends.[339]

The right to the support of a building is not a natural right; however, a right (easement) can be obtained by grant or prescription.[340] Where a building has enjoyed support from adjoining land as of right for more than 20 years, that factual support may well have been converted into a legal right or easement of support by prescription, where use over a lengthy period is recognised by the law as giving rise to a legal right for such use. If two buildings are in common ownership and one of them is sold to another party, this party will usually gain a right of support under section 62 of the LPA 1925. Where a neighbouring house has been demolished causing shrinkage to the clay beneath the foundation of the neighbour's house, this can be actionable.[341] Sections 38 (reiterated below for ease of reference) and 39 (relating to transitional provisions in First schedule) and Schedule 1 Part V paragraph 1 (set out below for reference) of the LPA 1925 refer to rights of support and of user in relation to party walls.[342]

Section 38 – Party structures

> *(1) Where under a disposition or other arrangement which, if a holding in undivided shares had been permissible, would have created a tenancy in common, a wall or other structure is or is expressed to be made a party wall or structure, that structure shall be and remain severed vertically as between the respective owners, and the owner of each part shall have such rights to support and user over the rest of the structure as may be requisite for conferring rights corresponding to those which would have subsisted if a valid tenancy in common had been created.*
>
> *(2) Any person interested may, in case of dispute, apply to the court for an order declaring the rights and interests under this section of the persons interested in any such party structure, and the court may make such order as it thinks fit.*
>
> …

335 *Gale on the Law of Easements* (n 11) 10-02.
336 *Redland Bricks Ltd v Morris* [1970] A.C. 652.
337 *Lotus Ltd v British Soda Co Ltd* [1972] Ch 123.
338 *Leakey v National Trust for Places of Historic Interest or Natural Beauty* [1980] Q.B. 485.
339 *Hunt v Peake* (1860) John. 705.
340 *Dalton v Henry Angus & Co* (1881) 6 App. Cas. 740.
341 *Brace v South East Regional Housing Association* [1984] 1 EGLR 144, CA.
342 *Gale on the Law of Easements* (n 11) 11-03.

Schedule 1 Part V paragraph 1

Where, immediately before the commencement of this Act, a party wall or other party structure is held in undivided shares, the ownership thereof shall be deemed to be severed vertically as between the respective owners, and the owner of each part shall have such rights to support and of user over the rest of the structure as may be requisite for conferring rights corresponding to those subsisting at the commencement of this Act.

A party wall built between two houses, which is the common boundary between these houses, is owned by both parties. Therefore, both parties enjoy an easement of support. Where support has been withdrawn, this is actionable only once damage occurs, not before.[343]

Section 2 of the PWA 1996 relates to work that can include acts that would be deemed to interfere with the right of support. Although carrying out such work is expressly authorised by the PWA 1996, liability to make good damage caused by the works is also expressly imposed by sections 2(3)(a), 2(4)(a), 2(5), 2(6) and 7(2) of the PWA 1996. Therefore, where the adjoining owner's right of support is impacted by work that has been carried out under the PWA 1996, such adjoining owner is protected by the above provisions of the PWA 1996. If a dispute arises under section 10 of the PWA 1996, an award may provide for works that need to be carried out to ensure adequate support to the adjoining owner's property. If the building owner fails to serve an appropriate party wall notice however, the PWA 1996 will not apply and the adjoining owner can go straight to court. Alternatively, if the disputants agree, they can agree on an ADR avenue. The question that arises here relates to the right of support and the effect of the works on such a right. An easement of support, whether part and parcel of the existence of a party wall or arising in respect of a boundary wall, continues regardless of whether the means of that support is altered.

There are two case examples relevant to the effect of section 9(a) of the PWA 1996, which states:

Nothing in this Act shall- (a) authorise any interference with an easement of light or other easement in or relating to a party wall; ...

Both of the cases pre-date the PWA 1996. The relevant provision at the time was as follows (showing that section 9(a) of the PWA 1996 is in sufficiently same terms):

Section 101 of the London Building Act 1894 states:

Nothing in this Act shall authorise any interference with an easement of light or other easements in or relating to a party wall or take away abridge or prejudicially affect any right of any person to preserve or restore any light or other thing in or connected with a party wall in case of the party wall being pulled down or rebuilt.

In *Selby v Whitbread & Co*,[344] the defendant was authorised by an award to demolish a building leading to removal of support from the flank wall of the defendant's building. This

343 *Midland Bank Plc v Bardgrove Property Services Ltd* (1993) 65 P. & C.R.153, CA; *Yorkshire Water Services Ltd v Sun Alliance and London Insurance Plc* [1998] Env. L.R. 204.

344 (n 6).

was followed by a second award, which required that the claimant supports the flank wall by erecting a pier of 'substantial character'.[345] However, the defendant failed to do so. The claimant sued for, amongst other matters, wrongful withdrawal of support.

McCardie J. held that the statutory rights superseded the common law right of support:[346]

> *The two sets of rights, namely, the rights at common law and the rights under the Act of 1894 (which followed the Act of 1855), are quite inconsistent with one another. The plaintiffs' common law rights are subject to the defendants' statutory rights. A new set of respective obligations has been introduced. The common law was seen to be insufficient for the adjustment of modern complex conditions. Hence I think that the Act of 1894 is not an addition to but in substitution for the common law with respect to matters which fall within the Act. It is a governing and exhaustive code, and the common law is by implication repealed.*

McCardie J. rejected the claim; however, he did grant damages to the claimant for the defendant's failure to build the pier. This makes sense as the first award authorised the support to be removed leading to the existing right of support being overridden. The second award was then upheld showing that the adjoining owner is protected. This is because the award secures adequate provision for continuing or restoring support which in turn protected the former right of support. As the pier was not erected, there was no question over whether the claimant had a right of support derived from the pier. This case dates from before the PWA 1996; however, it is relevant in terms of how a court can approach a schism between common law and statute law.

The second case is that of *Sack v Jones*.[347] The parties owned adjoining houses, which were separated by a party wall and had implied mutual rights of support. According to the claimant, the defendant's house was subsiding, dragging the party wall over, resulting in damage to the claimant's house due to lack of repair and underpinning. The court held that the claimant's allegations lacked evidence and even if there was sufficient evidence, there would have been no cause of action.

The court (Astbury J.) held:[348]

> *Now although the defendant's house is subject to an easement of support in favour of the plaintiff's house, the defendant is under no obligation to the plaintiff to keep her own house in repair for that purpose.... Even if there were any pull in the present case it can only have arisen from subsidence of the defendant's house and flank wall in the ordinary course of nature without any action on the part of the defendant. The non-liability of the servient owner to repair appears also from Chauntler v. Robinson and Dalton v. Angus .... In the present case the defendant has done nothing at all. Her tenant resides*

345 ibid. 740.
346 ibid. 752.
347 [1925] Ch. 235.
348 ibid. 240–242.

> *in her house, which is in good repair. This action is brought owing to the belief of the plaintiff and her advisers that the defendant's house is pulling the plaintiff's over. That hypothesis has not been established, but if it had, I do not think that in the particular circumstances the defendant would be in any way liable. The action fails, and must be dismissed with costs.*

The interference with the easement of support was not caused by the defendant but by natural causes for which the defendant could not have been held liable. However, see decision in *Leakey v National Trust for Places of Historic Interest or Natural Beauty*,[349] (referred to further below).

A notable case in relation to the right of support in connection with a party wall is *Brace v South East Regional Housing Association*,[350] where the claimant and the first defendant owned two terraced houses. The first defendant wanted to demolish its house with no intention of rebuilding it. The relevant legislation at the time, the London Building Acts (Amendment) Act 1939, did not apply to the properties. The claimant and the first defendant nevertheless decided to appoint surveyors and sign a party structure agreement including a schedule of permitted works. The issue was that the first defendant's demolition of its house resulted in the soil (clay) drying out. The shrinkage of the soil caused subsidence damage to the claimant's house. As a result, the claimant claimed damages for negligence and wrongful removal of support against both the first defendant and its surveyor. The judge at first instance held that the defendants were not negligent. However, he also held that the defendants were liable for wrongful removal of support. Although the Court of Appeal dismissed the first defendant's appeal, it did not decide the surveyor's appeal.

The Court of Appeal held that the claimant's existing right of support did not depend on the agreement between the parties that allowed the defendant to demolish its house. The agreement itself was not an award (albeit it was modelled on an award under the London Building Acts (Amendment) Act 1939). In addition, the London Building Acts (Amendment) Act 1939 did not apply. The main point of the case was whether the claimant abandoned its right of support. As a matter of construction of the agreement, it was held that the claimant had not abandoned it. The decision is highly relevant to the right of support in the context of party walls, although it is silent on the issue contained in section 9(a) of the PWA 1996.

Case law is aligned with the PWA 1996 in that the PWA 1996 does not authorise interference with an easement as per section 9(a). This gives clarity and shows consistency in both legislative and case law developments. While party walls and related works benefit from certain freedoms granted by the PWA 1996, these are not meant to trump easements and linked rights and obligations of the relevant parties. The PWA 1996 both protects easements of support (in section 9) and also allows the means of that support to be altered (by authorising notifiable works, which do so).

---

349 (n 338).
350 (n 341).

6.3.2 Weather Protection

There is no easement of protection from the weather,[351] which essentially means that a building owner has the right to remove its neighbouring wall or building even if this leads to exposing its neighbour's external wall to the 'full blast of the elements'.[352] Weather protection is mentioned in the context of party walls here as it can often be linked to the easement of support.

According to *Gale on the Law of Easements*:[353]

> *In the light of this decision [Upjohn v Seymour Estates Ltd [1938] 1 All E.R. 614], it may be that the ruling in Phipps v Pears, in a case which did not involve a party-wall, that there is no easement to be protected from the weather, needs to be modified in the case of the kind of party-wall which is within s.38 of the LPA 1925. With such a party-wall, it can be said that the owner of one-half of the wall is entitled to use the other half of the wall, not only for support but also for protection from the weather and this right is conferred by the statute. The owner of one-half of a party-wall can in an action against a third party only recover damages in respect of the one-half vested in him.*

While weather protection as such is not an easement, the need to ensure that the adjoining owner has such protection makes sense in the context of the right to support, considering that long-term weather impact on an exposed party wall can cause such support to diminish or disappear completely as time progresses.

The building owner may be liable in damages at common law if it removes a wall leading to exposure of an internal wall for the first time or if it exposes an external wall or any other wall to serious wind suction.[354] In relation to damages of this kind, section 7(2) of the PWA 1996 provides for compensation and special precautions in relation to such damages, which should form part of the party wall award.

In *Marchant v Capital & Counties*,[355] as shown in Diagram 6.1, the claimant's new house was built against and had enclosed on the back wall of a warehouse owned by the defendant. The defendant then demolished the warehouse complying with the party wall rules valid at the time. The part used by the claimant's house was left erected in its original place.[356] The previously internal party wall dividing the claimant's premises from the demolished premises was left exposed to the weather. The surveyors' award imposed an obligation on the defendant to "*Maintain the exposed face of the party wall in a weatherproof condition.*"[357]

After a few years, rain penetrated through the exposed face of the wall resulting in damage to the claimant's property. As a consequence, the claimant sought to enforce the obligation against the defendant. The court held that the defendant was in breach of its continuing obligation contained in the surveyor's award to maintain the wall in a weatherproof condition. While this is a slightly unusual ruling, considering that continuing obligations are not imposed lightly by surveyors, imposing such a continuing obligation is within a surveyor's power.

---

351 *Phipps v Pears* [1965] 1 Q.B. 76.
352 Bickford-Smith, Nicholls and Smith (n 22) 15.11.
353 (n 11) 11-03.
354 Bickford-Smith, Nicholls and Smith (n 22) 15.11.
355 [1983] 1 WLUK 326.
356 Chynoweth (n 276) 266.
357 *Marchant v Capital & Counties* (n 355).

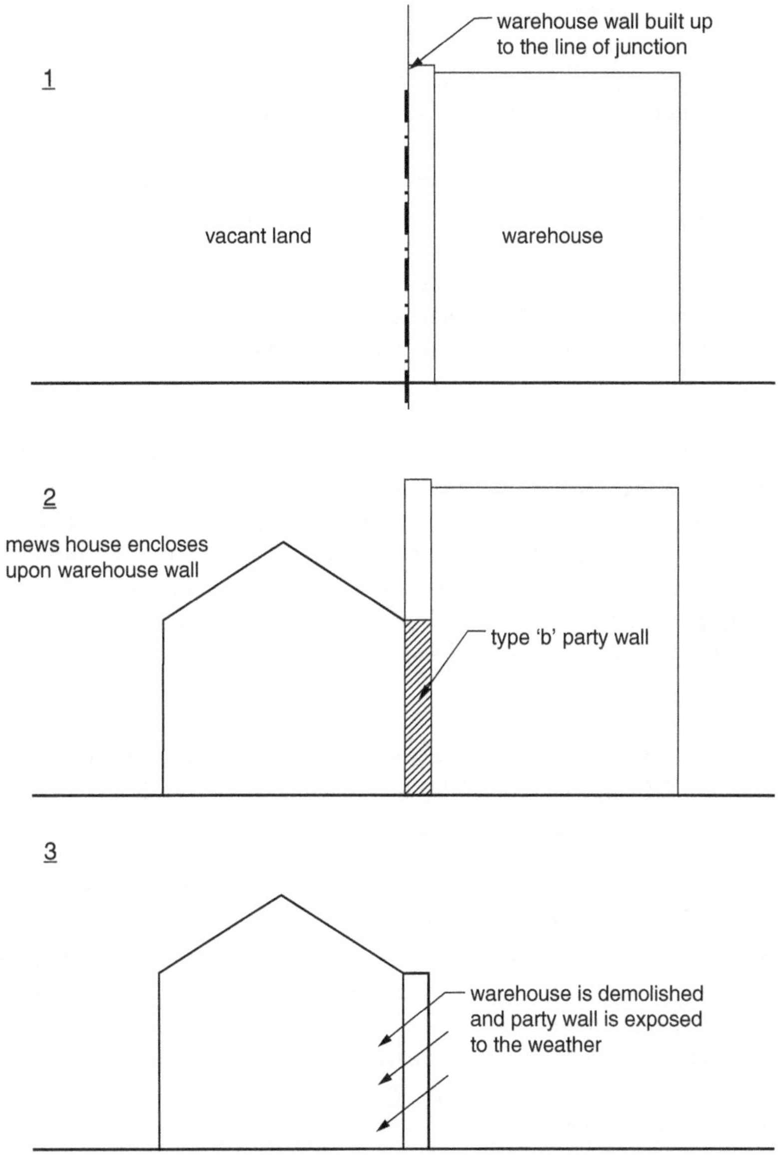

Diagram 6.1 *Marchant v Capital & Counties.*[358]

In *Rees v Skerrett*,[359] it was held that an award can validly impose an obligation to maintain a party wall on the building owner once works have been completed, meaning that the building owner is required to protect the wall from weather.[360] The Court of Appeal also held that a building owner who had demolished its property is liable for the damage caused

358 Chynoweth (n 276) 267.
359 [2001] 1 W.L.R. 1541.
360 Bickford-Smith, Nicholls and Smith (n 22) 15.11.

to a building of an adjoining owner caused by wind damage and water penetration. It also held that structural damage that is caused by wind could fall under the obligations related to the easement of support. As a result of the wind, it became clear that the wall on its own was not stable enough and required support. Although there is now a right in common law that protects from the weather, the court applied *Leakey v National Trust for Places of Historic Interest or Natural Beauty*,[361] imposing a positive duty on the landowner to "*do that which is reasonable in all the circumstances to prevent or minimise the known risk of damage to the neighbouring property.*"[362]

In *Rees v Skerrett*, the claimant recovered damages for the losses caused by both the withdrawal of support as well as by the weather damage. The withdrawal of support argument is the easement of support route, whereas the weather damage argument is the nuisance or negligence route. According to the court (Lloyd J.):[363]

> *Though it is not relevant on the facts to the present case, it is worth noting that the carrying out of works such as those undertaken by Mr Skerrett would now be governed by the provisions of the Party Wall etc Act 1996. By section 2(2)(n) of that Act a building owner has the right to expose a party wall hitherto enclosed subject to providing adequate weathering. Though an owner who is subject to a demolition notice does not have to serve a party structure notice under section 3 (see section 3(3)(b)) he is still liable to compensate his neighbour for loss or damage resulting from the work done under section 7.*

Extension of the common law by the decision of this case is not as revolutionary as it may have appeared at first considering that now the PWA 1996 would apply to a situation similar to the one in the above case.

The decision in *Phipps v Pears*[364] dates from 1965, prior to the PWA 1996 having been enacted. However, the decision in *Rees v Skerrett*[365] dates from 2001, after the PWA 1996 had been enacted pointing to the PWA 1996, which deals with weather-proofing in the context of party walls and states that when exposing a party wall, a building owner needs to provide adequate weather-proofing. This statutory principle promoted in *Rees v Skerrett*[366] is further supported by the above quotation from *Gale on the Law of Easements.*[367] This shows a departure from the approach in *Phipps v Pears*[368] as, while weather-proofing is not an easement as such, the potential hardship to an adjoining owner in the context of party walls and easements seems to have been addressed by the PWA 1996.

---

361 (n 338).
362 ibid. 524.
363 *Rees v Skerrett* (n 359), 34.
364 (n 351).
365 (n 359).
366 ibid.
367 (n 11) 11-03.
368 (n 351).

6.3.3 Excavations

Support can be either 'subjacent' (from below) or 'adjacent' (from the neighbouring property). Excavation works on the building owner's land are covered by section 6(10) of the PWA 1996, which provides that the building owner is not relieved from "*any liability to which he would otherwise be subject for injury to any adjoining owner or any adjoining occupier by reason of any work executed by him.*"

By way of example, in *Ann and Margaret Dodd v Samuel and James Holme*,[369] the defendant performed excavation works on his own soil. The claimant's building was in dilapidated condition and, as no shoring was provided, its building collapsed. The collapse was caused by the defendant's excavations. The issue in question was whether a party making such excavations is obliged to ensure that the adjoining owner's property's foundations are not weakened, and, should they be weakened, whether the party making the excavations is guilty of actionable negligence. The defendant was held liable in negligence.

Where the neighbouring land is to be excavated for the purposes of building a basement, the party excavating the land has the duty to provide adequate shoring or a strong retaining wall on its own land at its own cost. If, as the result of the excavation, the adjoining owner's land suffers damage, the adjoining owner is entitled to compensation by the building owner who is responsible for the natural support of the adjoining owner's land.

Another example of excavation works negatively affecting the adjoining owner's part of the party wall is shown in *Ray v Fairway Motors (Barnstable)*,[370] where the parties owned adjacent yards. The defendant performed excavation works in its yard down to a depth of four feet. This was immediately next to the claimant's boundary wall (Diagram 6.2). As a result, the claimant's wall and a shed subsequently built against it were damaged. The claimant sought to recover his losses from the defendant due to the defendant's alleged negligence and for breach of the claimant's natural rights and easements of support.[371] It was held that the defendant failed to prove that the shed placed on top of the party wall would justify the assertion that the easement of support was extinguished.

> *... the defendants failed to prove that the erection of the workshop produced such an additional burden on the servient tenement that the easement of support should be treated as extinguished. On this basis, the defendants did not seriously contend that they were not liable for the full amount of the damages awarded by the trial judge.*[372]

It was further held that the claimant's claim for damages for infringement to its easement succeeded; however, its alternative claim in negligence failed as an owner of land is entitled to do what they wish with their own land.[373]

---

369 110 E.R. 1296.
370 (1969) 20 P. & C.R. 261.
371 Chynoweth (n 276) 317.
372 *Ray v Fairway Motors (Barnstable)* (n 370), 274.
373 ibid.

Before starting excavation works falling under the PWA 1996 affecting a party wall, the building owner is advised to consult a chartered building surveyor or structural engineer to assess how the excavation works could affect the party wall and their long-term consequences. This will aid in assessing what (if any) excavation works are viable and communicating with the adjoining owner to agree on the best way forward. Such an agreement should be in writing for evidentiary purposes at a later point in time should a dispute arise.

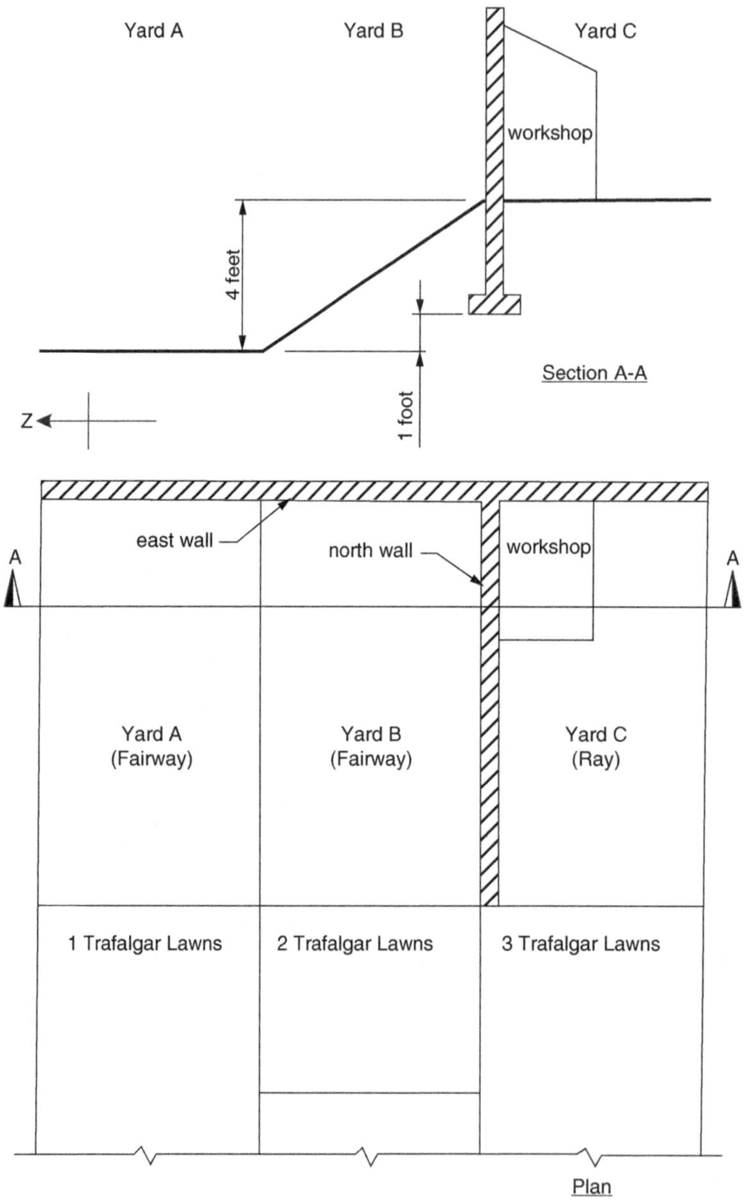

Diagram 6.2 *Ray v Fairway Motors (Barnstable).*[374]

374 Chynoweth (n 276) 319.

6.3.4 Easements – Right of Support – Conclusion

Where adjoining buildings are joined by a party wall, they mutually support each other. Where one of such buildings is demolished, it can cause serious issues leading to the collapse of the adjoining building. The owners of both buildings owe a duty of support to each other. Where a party decides to demolish its building, it must provide continued support to the remaining building on the adjoining owner's land. The right of support and related duty of support also applies where ground is removed too close to a neighbouring building. The same right of support applies also in situations where there is a risk that ground, as opposed to a building, will collapse due to excavation on neighbouring land. A right of support applies where retaining walls coincide with boundaries between properties.

The PWA 1996 applies to work carried out pursuant to section 2, which includes acts that could interfere with the right of support. The PWA 1996 authorises such work but also protects the affected adjoining owner (sections 2(3)(a), 2(4)(a), 2(5), 2(6) and 7(2) of the PWA 1996). Where a dispute arises, an award under section 10 of the PWA 1996 will address how such works should be performed to ensure the adjoining owner's property is adequately supported. This is subject to the building owner having served an appropriate party wall notice in time under the PWA 1996. Otherwise, if a dispute arises, parties would have to resort to court, unless they agreed on an ADR avenue.[375]

An owner of land who demolished its property is liable for the damage caused to an adjacent building by wind and water penetration. Excavation works on the building owner's land are covered by section 6(10) of the PWA 1996 under which the building owner is liable for injury to any adjoining owner or Adjoining Occupier resulting from work executed by the building owner.

The right of support is an easement that needs to be considered carefully before a party starts demolishing a building affecting a party wall or performing works that could negatively impact a party wall and related support to the adjacent building. Whether a party wishes to perform such works or is at the receiving end, it may be a good idea to consult a chartered building surveyor or structural engineer for advice before such works commence. This will ensure that the party/parties are well informed and can communicate before a dispute ensues as well as make sure that there is a record of any such communications and any relevant evidence in relation to such works.

6.4 Easements – Rights to Drainage

In *Pyer v Carter*,[376] the court held:

> *It seems in accordance with reason, that where the owner of two or more adjoining houses sells and conveys one of the houses to a purchaser, that such house in his hands should be entitled to the benefit of all the drains from his house, and subject to all the drains then necessarily used for the enjoyment of the adjoining house, and that without express reservation or grant, inasmuch as he purchases the house such as it is.*

375 Bickford-Smith, Nicholls and Smith (n 22) 15.10.
376 156 E.R. 1472, 1474.

Party walls, while separate from drainage rights, could become intertwined with such rights if drainage runs through them or is affected by them. In the case of *Walby v Walby*,[377] it was held that the test for implying an easement of necessity is a strict one:[378]

> *He accepted that the test for necessity was a strict one and that the facts must be such that the land retained cannot be used at all without the implication of an easement. Applying this approach to the present facts, Morgan J. concluded that the absence of a drainage easement did not prevent the retained land being used at all. Accordingly, it would not be right to imply the reservation of a drainage easement on the ground that it was an easement of necessity.*

It is not enough to show that the easement is necessary for the reasonable enjoyment of the land retained. In addition, the facts of the case need to show that the retained land cannot be used at all unless an easement is implied.

In the case of *Arena Property Services Ltd v Europa 2000 Ltd*,[379] a dispute arose in relation to a party wall dividing 98 and 96 Farringdon Road, EC1. The owners of the freehold of no. 98 (the defendant) and the owners of a leasehold in no. 96 (the claimant) planned to do work to the party wall and therefore served the relevant notices. Works performed by freeholders of no. 98 gave rise to a claim for damages in nuisance. Separately, the claimant removed a soil pipe, which served the defendant's property and protruded onto the claimant's property. The initial decision was that the claimant was not liable for compensation payment to the defendant. Subsequently, the defendant appealed the decision. The defendant argued it was entitled to compensation as an adjoining owner under section 7(2) of the PWA 1996 and that the soil pipe was covered by an easement. The claimant's argument was that the soil pipe amounted to trespass.

The relevant drainage pipe, which is at the crux of the dispute, ran through both no. 98 and no. 96 properties:[380]

> *The soil vent pipe came from 98 Farringdon Road and passed across the party wall at the first floor level and travelled to near the ground, at which point it passed through the party wall back on to the property forming part of 98 Farringdon Road. From there it was, or ought to have been, connected to the mains drainage. This soil vent pipe served the upper floors of 98 Farringdon Road which, as I shall explain, were subsequently the subject of a leasehold interest created in favour of the appellants in May 2001.*

The pipework attached to the outside of the building was in the vertical plane. One of the main questions was whether the surveyor's award was capable of extinguishing the easement (if there was such an easement). At first instance, it was held that the defendant failed to establish the existence of an easement and that a surveyors' award was capable of

377 [2013] 1 P. & C.R. DG6.
378 ibid. D14.
379 [2003] EWCA Civ 1943.
380 ibid 3.

extinguishing the easement (had there been one) as the PWA 1996 provided its own code, which supersedes the common law with the only easements being preserved noted in section 9 of the PWA 1996.

On appeal, the defendant tried to argue that, as the claimant relied on the PWA 1996, it had implicitly accepted that an easement existed. The appeal was dismissed. The Court of Appeal held that the surveyors' award would not have extinguished the easement. This could have been extinguished only when the claimant rendered the pipe unusable. However, the Court of Appeal agreed with the first instance judge on the point that the defendant failed to establish that an easement existed for more than twenty years. In fact, the easement was not extinguished until the claimant blocked the pipe so that it could not serve any use for the adjoining owner (the defendant).

> *... s.9 would seem to exclude the possibility of any works which might permanently deprive an adjoining owner of his easement. On this basis, Arena would have had no right to remove the soil pipe without simultaneously preserving the substance of the easement. In practice, this would have required the pipe to be re-routed, as was previously the practice at common law as well as under the London Building Acts. Indeed, as has already been noted, this is now standard practice under s.2(5) of the 1996 Act. In these circumstances, the dubious possibility of buying out an easement with compensation under s.7(2) should never need to arise.*[381]

The case shows that drainage rights are capable of being an easement and can become a contentious matter in connection with party walls. It is not the intention of the PWA 1996 to interfere with easements which is clear from the wording of section 9(1) of the PWA 1996.

As a conclusion to drainage rights, a right of drainage can transpire independently from a party wall if, for example, it runs through or along the party wall, in which case it will have to be taken into account.[382] Drainage rights are relevant to party walls and should be considered carefully if a potential interference between the two is a possibility. The PWA 1996 is not meant to 'trump' easements, and therefore an award will not be capable of extinguishing an easement on its own.

6.5 Linked to Easements – Are There Rights or Obligations to Repair a Party Wall?

6.5.1 Party Wall etc. Act 1996 and Repairing a Party Wall

When it comes to party walls, naturally they require upkeep once built. A right of access for the purposes of repair is an easement. The right/obligation to repair a party wall is not an easement. The question arises, who has the obligation to repair a party wall (if such an obligation exists at all) and also whether one of the parties relevant to the party wall can assert a right for the other party to repair the wall.

---

381 P Chynoweth, 'The Party Wall etc Act 1996: compensation and treatment of easements' (2004), *The Conveyancer and Property Lawyer*, 240.

382 C Cant, 'Easements – Adjoining properties and party walls' (2013) https://www.christophercant.co.uk/wp-content/uploads/2020/09/Easements-Adjoining-properties-and-party-walls.pdf.

According to *Megarry & Wade*,[383] neither of the parties relevant to the party wall are obliged to repair their own half of the party wall. However, each has the right to repair the neighbour's half, which is a matter of choice/free will.[384] In addition, neither of the landowners or occupiers have the right to pull down the party wall (with the exception of rebuilding it with reasonable dispatch) or to demolish their own half of the party wall meaning removal of the other party's support.[385] Also, neither of the parties is allowed to prevent the other party from enjoying a section of the party wall.[386]

Where an existing party wall falls into disrepair and/or is damaged and either the building owner or the adjoining owner wants to ensure that this is addressed against the other, there are ways to compel the other party to do so. Although it is not possible to force the other party to cooperate in repairing the party wall under the PWA 1996 procedure directly, it is possible to obtain a party wall award. The award can authorise the building owner to exercise its right to repair the party wall and to recover an appropriate contribution from the adjoining owner to the cost of the work. The award gives some comfort in that it provides protection if the other party tries to complain later down the line about works being done on the party wall. Another advantage of such an award is also that it can allocate costs for the works done on the party wall where such works are for the purposes of repairing existing structures.[387]

Section 11(1) states that expenses of work under the PWA 1996 are usually defrayed by the building owner (with exceptions set out in section 11 of the PWA 1996). Should a dispute arise, it is possible under the PWA 1996 to obtain a party wall award (subject to the building owner having served an appropriate party wall notice in time).

According to DCLG's Explanatory Booklet, works that are small in nature, such as simple repairs (for example, re-plastering or cutting into the party wall to add or replace electrical wiring and sockets), do not require service of notice under the PWA 1996.[388] One could however argue that it is not clear why it should be possible to cut chases into the party wall, which have considerable potential to damage decorative finishes on an adjoining owner's property, without having to serve notice when merely cutting a slot for flashing would require service of a notice.

While *de minimis* works to a party wall are not of sufficient size or impact that trigger the operation of the PWA 1996, for larger/significant works to a party wall, the parties can simply agree amongst themselves what needs to be done, which should be recorded in a party wall agreement (which is not a party wall award and does not require the involvement of a party wall surveyor). The party wall agreement will ensure clarity of what the parties agreed amongst themselves and what works have been performed. This is a positive step the parties can take to mitigate the likelihood of a dispute arising and at

---

383 Megarry & Wade (n 8) 29-047.

384 *Jones v Pritchard* [1908] 1 Ch. 630; *Sack v Jones* (n 347).

385 *Brace v South East Regional Housing Association* (n 341); *Bradburn v Lindsay* [1983] 2 All E.R. 408.

386 *Dewdney Stedman and Smith Stedman v Henry Attwell Smith* (1857) 8 El. & Bl. 1.

387 R Pratt, '4 ways you might be affected by the Party Wall Act' (LexisNexis, 2013), https://www.lexisnexis.co.uk/blog/dispute-resolution/4-ways-the-party-wall-act-might-affect-you.

388 (n 26) 6.

the same time protect themselves should a dispute arise by having a clear record of what the parties have agreed, which the court will take into account should it come to litigation.

Section 2 of the PWA 1996 covers 'repair etc. of party wall: rights of owner'. Works falling under section 2 of the PWA 1996 require service of a notice by the building owner under section 3 ('party structure notices') of the PWA 1996 on every adjoining owner in relation to neighbouring properties that are affected by such works at least two months prior the works begin. The affected adjoining owner has the right to agree or disagree in writing or do nothing. If the adjoining owner chooses to do nothing for a period of at least fourteen days, a dispute is deemed to have arisen and will need to be dealt with by a party wall surveyor. If the adjoining owner disagrees expressly, an actual dispute will arise. For avoidance of doubt, and as noted earlier, it is not the intention for the exercise of statutory rights to authorise interferences with any easements connected with party walls.[389]

6.5.2 Easements and the Right vs. Obligation to Repair

As for easements, these relate to rights to "*make some limited use of land belonging to somebody else. An example is a private right of way over a neighbour's* land."[390] According to *Gale on the Law of Easements*: "*the question of ancillary easements commonly arises in connection with repairs.*"[391] This becomes relevant to the area of party walls where a party wall becomes an inherent part of an easement, for example, where it is the subject matter of the right of way or right to support. In this context, distinction must be made between the 'right to repair' and the 'obligation to repair'.

6.5.2.1 Is There a Right to Repair?

In *Jones v Pritchard*,[392] Parker J. held:

> *Once again, the grant of an easement is prima facie also the grant of such ancillary rights as are reasonably necessary to its exercise or enjoyment. Thus the grantee of an easement for a watercourse through his neighbour's land may, when reasonably necessary, enter his neighbour's land for the purpose of repairing, and may repair, such watercourse. On this principle each party in the present case may do such acts on the property of the other as are reasonably necessary to the continued enjoyment of the easement; for example, each party would be entitled to repair the other's half of the wall in question so far as was reasonably necessary for the enjoyment of any easement impliedly granted or reserved: see Pomfret v. Ricroft.*[393]

Where the person entitled to the easement carries out repairs on the servient tenement to prevent damage occurring to the servient property, it is that person who has to bear the burden of the repair costs and is not able to claim a contribution towards such costs from the owner

389 PWA 1996, s 9(a); *Crofts v Haldane* (1866-67) L.R. 2 Q.B. 194.
390 Law Commission – Ministry of Justice, 'Impact Assessment of the Law of Property Bill [2011]' (2011).
391 (n 11) 1–112.
392 (n 384) 638.
393 ibid. 638.

of the servient tenement. There is no obligation on the servient owner to repair the subject matter of an easement and therefore is allowed to let it fall into disrepair.[394] This issue will likely be the subject to the actual terms of any grant or what the parties agree between themselves by way of a contract. The position changes if a nuisance needs to be remedied and case law has an impact as well when it comes to the question of sharing of costs[395] between the affected parties.

In *Pomfret v Ricroft*, it was held:[396]

> *As if a man gives me a license to lay pipes of lead in his land to convey water to my cistern, I may afterwards enter and dig the land to mend the pipes, though the soil belongs to another and not to me.*

Another example includes *Bond v Nottingham Corp*,[397] where the defendant (local authority) wanted to demolish a building under a statutory power. The issue was that the building was subject to an easement of support in relation of an adjoining building. The Court of Appeal held that, while an easement of support was in existence, this in itself did not impose any repairing obligations on the servient owner (even where the deterioration of the servient building meant that the dominant building was deprived of support). It was however also held that, where the servient owner demolishes its building, it is obliged to give equivalent support to the dominant building and that was the situation that the defendant was in when demolishing under a statutory power.[398]

> *The nature of the right of support is not open to dispute. The owner of the servient tenement is under no obligation to repair that part of his building which provides support for his neighbour. He can let it fall into decay. If it does so, and support is removed, the owner of the dominant tenement has no cause for complaint. On the other hand, the owner of the dominant tenement is not bound to sit by and watch the gradual deterioration of the support constituted by his neighbour's building. He is entitled to enter and take the necessary steps to ensure that the support continues by effecting repairs, and so forth, to the part of the building which gives the support. But what the owner of the servient tenement is not entitled to do is, by an act of his own, to remove the support without providing an equivalent. There is the qualification upon his ownership of his own building that he is bound to deal with it, and can only deal with it, subject to the rights in it which are vested in his neighbour.*[399]

The case shows that the right of support in itself does not create an obligation on the owner of the servient tenement to repair and does not oblige the owner of the servient tenement to prevent the building providing support to fall into decay. However, at the same time, the owner of the servient tenement must not remove the support without ensuring that it substitutes it with an equivalent to support the dominant tenement. There is a clear divide

---

394 Chynoweth (n 276) 194.
395 *Abbahall Ltd v Smee* [2002] EWCA Civ 1831.
396 85 E.R. 454, 459.
397 [1940] Ch. 429.
398 Chynoweth (n 276) 61.
399 *Bond v Nottingham Corporation* (n 397), 438, 439.

between the owner of the servient tenement acting passively or actively when it comes to its tenement providing support. The owner of the dominant tenement is not obliged to repair either. However, while the owner of the dominant tenement is entitled to enter the servient owner's land for the purpose of repair, the work performed has to be only that which is necessary and in a reasonable manner.[400] Where however the repair is not necessary, no right of entry on the servient tenement in order to perform the repairs will be implied.[401]

According to *Gale on the Law of Easements*:[402]

> *The dominant owner is entitled to enter the servient land to effect repairs or to alter the surface of the servient land to accommodate the right granted. So a dominant owner may be entitled to enter the servient land to install lighting if that is reasonably necessary to enable the way to be used safely and conveniently.*
>
> *The following summary of the law has been described as "settled for centuries" and uncontroversial. Subject to contrary agreement,*
>
> *(1) the grantee may enter the grantor's land for the purpose of making the grant of the right of way effective viz. to construct a way which is suitable for the right granted to him;*
> *(2) once the way exists, the servient owner is under no obligation to maintain or repair it;*
> *(3) similarly, the dominant owner has no obligation to maintain or repair the way;*
> *(4) the servient owner (who owns the land over which the way passes) can maintain and repair the way, if he chooses;*
> *(5) the dominant owner (in whose interest it is that the way be kept in good repair) is entitled to maintain and repair the way and, if he wants the way to be kept in repair, must himself bear the cost;*
> *(6) he has a right to enter the servient owner's land for the purpose but only to do necessary work in a reasonable manner.*

This was confirmed in *Regency Villas Title Ltd v Diamond Resorts (Europe)*:[403]

> *... step-in rights are, by definition, rights to reasonable access for maintenance of the servient tenement, sufficient, but no more than sufficient, to enable the rights granted to be used: see Gale on Easements, 20th ed, para 1-93 and Carter v Cole [2006] EWCA Civ 398; [2006] NPC 46 per Longmore LJ. The dominant owner's right is "to enter the servient owner's land for the purpose, but only to do necessary work in a reasonable manner ..."*

Where, on the other hand, the owner of the dominant tenement agrees to contribute to the cost that the servient tenement owner incurred when performing repairs to a road, the servient owner cannot charge the dominant owner above and beyond works which would have been done purely to repair (e.g. the road) in line with the covenant.[404]

---

400 *Carter v Cole* [2006] EWCA Civ 398.
401 *Gale on the Law of Easements* (n 11) 1-112; *Edwards v Kumarasamy* [2016] UKSC 40.
402 (n 11) 9–112 to 9–113.
403 [2018] 3 W.L.R. 1603, 65.
404 *Gale on the Law of Easements* (n 11) 9–114.

A distinction needs to be made between express grants and prescriptive rights. According to *Gale on the Law of Easements*, where, for example, a right of way is granted for 'all purposes', then the owner of the dominant tenement can make improvements even for purposes that have not been contemplated at the time of the grant. On the other hand, a prescriptive right is restricted by the type of the use, which gave rise to it. Therefore, no improvements can be made to increase the burden on the servient tenement.[405]

In *Carter v Cole*,[406] Lord Justice Longmore noted:

> *The common law has contemplated with equanimity the prospect that both the servient owner and the dominant owner have the right of repair.*

While the owner of the dominant tenement has a right to enter and repair, it has also the right to prevent a substantial interference with such dominant owner's right.[407]

As for the extent of the ancillary right to repair, in *Edwards v Kumarasamy*,[408] Lord Neuberger held:

> *First, a right of way does not necessarily carry with it a right to carry out repairs to the way: such an ancillary right only arises as a matter of implication, and is normally justified because the servient owner has no obligation to repair the way. As it is put in Gale on the Law of Easements 19th ed (2012), para 1-90, "[t]he ancillary right arises because it is necessary for the enjoyment of the right expressly granted." In the present case, the Headlease, under which Mr Kumarasamy was granted the right to use the common parts, contains an obligation on the freeholder to keep the common parts in repair. Accordingly, I do not consider that it would be appropriate to imply such an ancillary right: it is not necessary for business efficacy, nor is it obvious.*

Where there is a covenant obliging a lessor to repair, which benefits the lessee, the courts will not imply a right to repair for the lessee.

When considering the principle that the owner of the servient tenement has no obligation to keep the servient tenement in good repair, while the owner of the dominant tenement is entitled to repair the servient tenement (but is not obliged to do so) with the accompanying right to enter the servient tenement – it is important to remember that the 'right to repair' needs to be examined and established in each individual case taking into account the surrounding circumstances as it is not an automatic right in itself.[409]

The Law Commission issued a report in 2011,[410] the recommendations in which include "*to make it possible for the benefit and burden of positive obligations to be enforced by and against subsequent owners.*"[411] This may result in greater clarity as to when the dominant owner has the right to 'step in' and repair when the burdened land changes owners. It is yet to be seen how the Law of Property Bill will address this issue.

---

405 ibid. 1–112.
406 (n 400) 19.
407 *Gale on the Law of Easements* (n 11) 1–113.
408 (n 401) 57.
409 *Gale on the Law of Easements* (n 11) 1–112.
410 Law Commission, 'Making Land Work: Easements, Covenants and Profits a Prendre' (2011) 327, https://www.lawcom.gov.uk/project/easements-covenants-and-profits-a-prendre/.
411 ibid.

A case that discusses the 'right to repair' specifically in relation to party walls is *Barry v Minturn*,[412] where damp penetrated from the exposed face of a type A party wall into the defendant's building as illustrated in Diagram 6.3 below. The defendant wanted to get rid of the damp that was in the wall and therefore suggested that a damp-proof barrier be erected on the claimant's side of the wall. The defendant served a notice to be able to repair the wall. It was held that the wall, which used to be a garden wall, continued to be a garden wall on the claimant's side of the property. It was further held that the wall was defective as it allowed damp to penetrate into a building. The defendant had a right therefore to repair the wall. The method of repair had to be selected by surveyors in order to not inconvenience the claimant unnecessarily. The defendant had no right to erect the damp-proof barrier on the claimant's side. Equally, the surveyors had no authority to require the defendant to erect the damp-proof barrier on her side (which she was, however, free to do, should she wish to do so). However, the defendant was held to be entitled to insert the barrier within the thickness of the wall.[413]

The issue was that the award required that a specific repair solution be undertaken to a party wall (partly due to the party wall's prior history). The House of Lords considered this to be an improper basis for a decision. Nevertheless, it was not willing to overturn the award given that the evidence showed that the tribunal making the award took the level of inconvenience caused to the adjoining owner into consideration, as was required by statute.[414]

Lord Parker of Waddington stated:[415]

> *I am not satisfied that it was, on the true construction of the Act,*[416] *open to the County Court, at any rate under the particular circumstances of this case, to make any award which would involve the permanent appropriation of any land either of the appellant or the respondent. I am satisfied that it was not open to him to direct any works which would entail great inconvenience to the appellant, if it were possible to direct other works which, while equally effective and not involving any considerable extra cost so far as the respondent was concerned, would be accompanied by no such inconvenience. The County Court judge adopted, and I think rightly adopted, the only other alternative open to him on the evidence by directing the works mentioned in his award, which works he finds, as a fact, will be perfectly effective, and can be carried out without entry on the appellant's land.*

Lord Parker of Waddington also noted:

> *In my opinion, dampness in a wall is not a defect within the meaning of the Act unless its existence renders the wall less effective for the purposes for which it is used or intended to be used. If a wall between two gardens is damp, it is not on that account defective, for its dampness is immaterial ... The County Court judge has found that the only defect in the wall is that part of it allows damp to percolate into the respondent's premises, and that this defect will be effectively remedied by the work he directed. This finding as a finding of fact is conclusive between the parties, and I do not think that it involves any error of law.*[417]

---

412 [1913] AC 584.

413 Chynoweth (n 276) 52.

414 Paul Chynoweth, 'Invalid party wall awards and how to avoid them' (University of Salford Manchester 2000), 14, https://salford-repository.worktribe.com/output/1457272/invalid-party-wall-awards-and-how-to-avoid-them.

415 *Barry v Minturn* (n 412) 595.

416 The 'Act' Lord Parker refers to is the London Building Act 1894.

417 *Barry v Minturn* (n 412) 594.

While a party cannot intrude onto the other party's side of the property, courts will be careful to ensure that, where there is a defect to a party wall that can be effectively remedied, the party affected by the defect will be entitled to perform the relevant repairs (here insert the damp-proof barrier within the thickness of the party wall) to such an extent that the repairs do not unnecessarily inconvenience the other party.

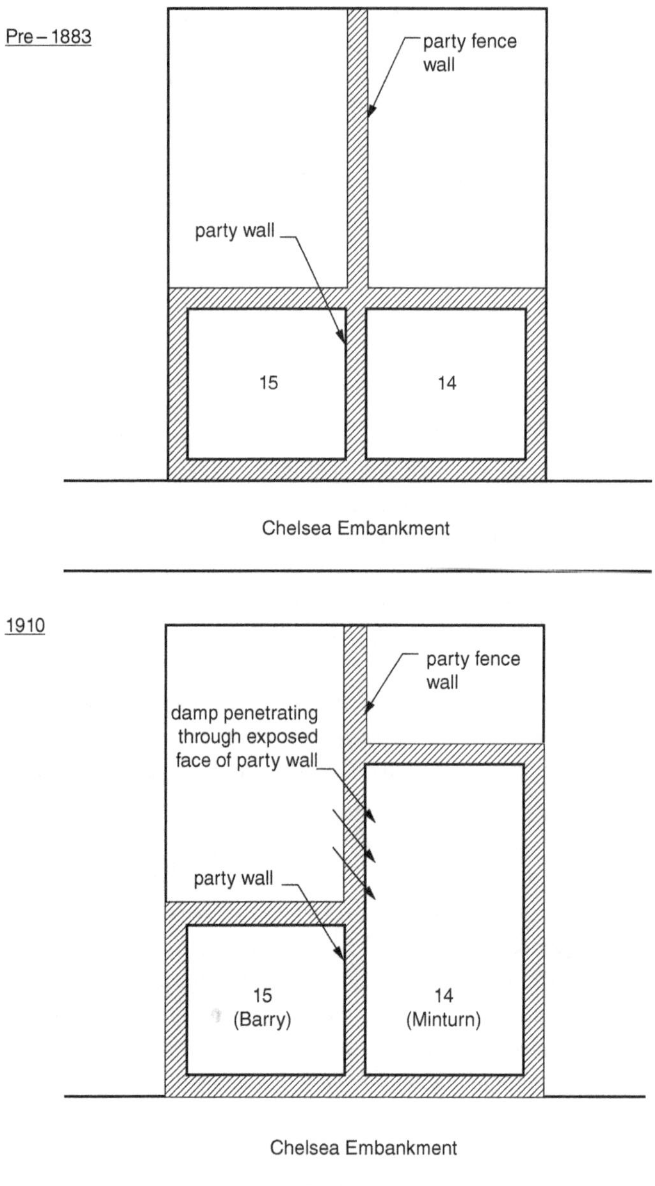

Diagram 6.3 *Barry v Minturn.*[418]

418 Chynoweth (n 276) 53.

6.5.2.2 Is There an Obligation to Repair?

In comparison to the 'right to repair', the 'obligation to repair' refers to a duty or the burden of the responsibility for the dominant owner to keep the subject of an easement in repair. The general principle is that the dominant owner has no duty to keep the subject of its easement repaired (as noted above).[419] However, this is a generalised statement and is subject to a number of exceptions. The fact remains that, similar to the 'right to repair', whether there is an 'obligation to repair' depends on the circumstances of each individual case. While, technically, there is no principle of the 'obligation to repair', below are examples of practical situations where an 'obligation to repair' on the part of the owner of the dominant tenement may arise.

In *Taylor v Whitehead*,[420] Lord Mansfield stated that: "*by common law, he who has the use of a thing ought to repair it.*" According to *Gale on the Law of Easements*, this would mean that the owner of the servient tenement has no obligation to repair and that, therefore, if the owner of the dominant tenement wants, for example, the right of way repaired, then it has to perform the repair works itself. While the dominant owner cannot be forced to repair, in practical terms, it has no other choice in order to ensure that it is not committing trespass or nuisance.[421] This is in the context of the obligation not to cause nuisance, for example, where smoke or water is encroaching onto the servient tenement.[422]

This is further supported by the decision in *Ingram v Morecraft*,[423] where Sir John Romilly M.R., albeit obiter, notes:

> *If I grant a man a right to lay pipes over my land, it follows that he must keep them watertight, for otherwise the escape of water is trespass.*

In *Jones v Pritchard*,[424] Parker J. elaborated on the issue:

> *... there is undoubtedly a class of cases in which the nature of the easement is such that the owner of the dominant tenement not only has the right to repair the subject of the easement, but may be liable to the owner of the servient tenement for damages due to any want of repair. Thus, if the easement be to take water in pipes across another man's land and pipes are laid by the owner of the dominant tenement and fall into disrepair, so that water escapes on to the servient tenement, the owner of the dominant tenement will be liable for damage done by such water. Strictly speaking, I do not think that even in this case the dominant owner can be said to be under any duty to repair. I think the true position is that he cannot, under the circumstances mentioned, plead the easement as justifying what would otherwise be a trespass, because the easement is not, in fact, being fairly or properly exercised.*

---

419 *Gale on the Law of Easements* (n 11) 1–114.
420 (1781) 2 Doug. K.B. 745, 749.
421 (n 11) 1–114.
422 *Jones v Pritchard* (n 384).
423 (1863) 33 Beav. 49, 51–52.
424 (n 384) 638.

According to *Gale on the Law of Easements*, it does not make a difference that the pipe serves not only the dominant tenement but also the servient tenement[425] (who has the right to repair but not an obligation to do so). The main distinguishing feature between the 'right to repair' and the 'obligation to repair' is that while for either to be established, circumstances of each individual case need to be considered, the 'right to repair' is a possibility or an opportunity for the dominant owner to perform repair works on the servient tenement (therefore accessing it) in order to be able to enjoy the easement (not being prevented from carrying out such works by the servient owner) as long as the works are 'reasonable'. The practical concept of an 'obligation to repair' arises where the dominant owner runs the risk of trespass or nuisance if it does not perform the repair works.[426]

Both issues, the 'right' and the 'obligation' to repair, are relevant to party walls as can be seen in case law, including in *Barry v Minturn*.[427] Party walls are in place usually with an ancillary easement of support. Under common law, such an easement does not usually, with exceptions (as discussed above), impose an 'obligation to repair'. However, if nuisance arises due to neglect or failure to act, indirectly, an 'obligation to repair' may arise in practical terms, again depending on the circumstances of each particular case. An adjoining owner who suffers from the lack of repair can then try to recover damages to cover costs it incurred as a result of performing the repair works on the party wall itself. Interestingly, where a party wall results in being exposed to the elements due to disrepair, the damages can also cover the cost of finishing such a party wall as to its appearance and reasonable standard of weather proofing. It is important to remember that where the PWA 1996 applies, this supersedes common law remedies when it comes to rights of support.[428] It is recommended to consider including an obligation to repair in the express grant of an easement (which is the usual source of obligations to repair).

Below are examples of case law, where an 'obligation to repair' or damages arose in relation to repair works to party walls.

In *Bradburn v Lindsay*,[429] dry rot in the party wall was caused by dilapidation and neglect in one of the semi-detached houses, also resulting in the local authority having to demolish the derelict house. This led to the party wall being unsupported to a large extent. As the defaulting party owed a duty to the adjoining owner to take reasonable steps to abate reasonably foreseeable nuisance, the adjoining owner was able to recover damages for covering the cost of getting rid of the dry rot and erecting the supporting buttresses.

Case law shows the courts' proclivity towards awarding damages where the building owner jeopardises the adjoining's owner's right of support by carrying out works to the relevant party wall.

In *Brace v South East Regional Housing Association*,[430] an end of terrace property was demolished. Shrinkage and subsidence were caused by the conversion of a party wall into

425 (n 11) 1–114.
426 ibid. 6–101.
427 (n 412).
428 Halsbury's Laws of England Online (n 254) 374.
429 (n 385).
430 (n 341).

a flank wall. As the right of support was interfered with causing nuisance, damages were recovered. This is even though a party wall agreement was in place.

In *Rees v Skerrett*,[431] again, an end of terrace property was demolished. On this occasion, damage from wind suction was caused by the conversion of a party wall into a flank wall. Damages were recovered as the right of support/protection covered the effect of the weathering in question. The damages related to the withdrawal of support as well as the breach of duty for the party wall to be weatherproofed once the demolition was completed:[432]

> *Indeed, the withdrawal of support where an easement of support exists may impose a duty to take reasonable steps to weatherproof a dividing wall once it is exposed to the elements as a result of the demolition ...*

This is given statutory effect by section 2(2)(n) of the PWA 1996.

However, damages in the examples above were recoverable only once the damage occurred, and these cannot be recovered in relation to expenses covering works for the purposes of preventing damage that is anticipated in the future.[433]

6.5.3 Covenants to Repair

The owner of a servient tenement is not obliged to perform repairs that are needed for the owner of the dominant tenement to enjoy an easement, apart from as required by statute or any special local custom or express contract. As discussed above, where a house falls into disrepair causing damage to an adjoining house that has the right of support, the adjoining owner may be entitled to damages.[434]

While the parties are free to enter into a covenant where the owner of the servient tenement covenants to keep in repair a perpetual easement, successors in title will usually not be bound by this (subject to an express assignment of the benefit of the covenant). This is subject to exceptions and has also been a topic of substantial debate.[435] For example, see the principle formulated in *Halsall v Brizell*[436] which obliges a dominant owner exercising a right over servient land to comply with any corresponding positive obligations, i.e. that a party may not take the benefit without accepting the burden that goes with it.

Gale on the Law of Easements states:[437]

> *It seems clear on principle that if a servient owner covenants to keep in repair the subject of a perpetual easement, the covenant, being a positive covenant, cannot be enforced against a successor in title.*

---

431 (n 359).
432 Halsbury's Laws of England Online (n 254) 374.
433 *Midland Bank plc v Bardgrove Property Services Ltd* (n 343); Halsbury's Laws of England Online (n 254) 374.
434 *Gale on the Law of Easements* (n 11) 1–115.
435 *Gaw v Coras Iompair Eireann* [1953] 1 WLUK 125; *Grant v Edmondson* [1931] 1 Ch. 1; *Gale on the Law of Easements* (n 11) 1–117.
436 [1957] Ch 169.
437 (n 11) 1–117.

Further, according to *Gale on the Law of Easements*:[438]

> *The burden and benefit of a covenant entered into by the grantor of an easement for a term of years devolves, no doubt, in accordance with the rules applicable as between landlord and tenant.*

Leasehold covenants are, however, a separate area from that of easements. For the purposes of the PWA 1996 (section 20), both the leaseholder and freeholder are viewed as 'owners'. Both the landlord and the tenant can covenant to do repair work in the leasehold agreement. A tenant can covenant to keep the premises 'in repair', which includes an obligation to put the property into repair where the property is in disrepair at the time the lease is entered into.[439] The landlord can enter into a repairing covenant where it is obliged or has the right to carry out repair work as is reasonably required to remedy defects at the property.[440] As noted in *Gale on the Law of Easements*, the benefit and burden of a leasehold covenant will run with the land transferred (as opposed to freehold covenants, which are positive in character, which will in general not run with the land transferred).[441] While a 'right' or 'obligation' to repair is possible in leasehold covenants, and therefore can extend to party walls, this is separate from easements or the PWA 1996.

6.5.4 Are There Rights or Obligations to Repair a Party Wall? – Conclusion

While it is not possible to force the other party to repair a party wall under the PWA 1996, it is possible to obtain a party wall award, which can allocate costs for the works done on the party wall if these are performed to repair existing structures.

The owner of the servient tenement has no obligation to keep the servient tenement in good repair. The owner of the dominant tenement is entitled to repair the servient tenement (but is not obliged to do so) with the accompanying right to enter the servient tenement.

In order to establish whether there is a 'right to repair' or a practical 'obligation to repair', circumstances of each individual case need to be considered. The 'right to repair' is a possibility or an opportunity for the dominant owner to perform repair works on the servient tenement (therefore accessing it) in order to be able to enjoy the easement (not being prevented from carrying out such works by the servient owner) as long as the works are 'reasonable'.

The 'obligation to repair' in practical terms arises where the dominant owner runs the risk of becoming responsible for trespass or nuisance if it does not perform the repair works.

While the parties are free to enter into a freehold covenant, where the owner of the servient tenement covenants to keep in repair a perpetual easement, successors in title will usually not be bound by this (subject to an express assignment of the benefit of the covenant).

The benefit and burden of a leasehold covenant will run with the land transferred (as opposed to freehold covenants). While a 'right' or 'obligation' to repair is possible in

---

438 ibid. 1–118.

439 *Proudfoot v Hart* (1890) 25 Q.B.D. 42.

440 M Richards, N Roberts, R Abbey, S Desmond, 'When is the covenantor liable to carry out repairs under his repairing covenant?' (1996) *Butterworths Property Law Service*, 804.

441 (n 11) 4–55.

leasehold covenants and therefore can extend to party walls, this is separate from easements or the PWA 1996.

Where a party wall falls into disrepair, parties should examine the nature of any potential easements that are intertwined with the relevant party wall and establish whether, in the specific individual case, a right/obligation to repair arises and by whom. Where a party wall falls into disrepair, either owner will have the right to repair under the PWA 1996, together with a potential right to recover a contribution from the other owner. In addition, there may be obligations to repair imposed by express covenant, implied covenant or obligation, or as a consequence of an associated prescriptive right.

6.6 Conclusion

The PWA 1996 encompasses rights that can affect a number of other areas of law in relation to the building owner's and adjoining owner's relationship as to party walls. It is not the intention that the PWA 1996 affects other rights except where it is necessary to give effect to the purposes of the PWA 1996.[442]

Easements are one of the areas or legal rights that can be intertwined with the topic of party walls. Right of way, right of support, drainage rights and right of light are the most obvious easements that are relevant to party walls.

The building owner and the adjoining owner need to consider easements and check if any ancillary rights and/or obligations arise in relation to the relevant party wall. This is to ensure that they do not fall short of their obligations and that they can enjoy their rights as well. Section 9(a) of the PWA 1996 is clear in that it states that the PWA 1996 does not authorise "*any interference with an easement of light or other easements in or relating to a party wall ...*"[443]

The easements discussed in the context of party walls here include rights of way, rights of support, drainage rights, and the discussion on the ancillary right or obligation to repair (which are not easements), with the right to light addressed later in the context of Chapter VII.

Easements discussed in this chapter can interact with party walls or co-exist with them; however, such easements are not limited to those discussed in this book. Those discussed in this book are easements that appear in practice most often in the context of party walls and therefore should be considered carefully; however, parties should look beyond those easements set out in this book with the right legal advice in place.

A party wall award under the PWA 1996 is unlikely to be successful in achieving a change in an easement, whether it is attempting to create or extinguish one.

As for the 'right to repair' and practical 'obligation to repair', while these are not easements themselves, a discussion on this topic is included in this chapter as they can be linked to easements and can become highly relevant in the context of party walls.

While party walls are to a large extent governed by the PWA 1996, the fact remains that the PWA 1996 is not an isolated system of rights and obligations in relation to party walls, which are affected by a number of other areas of law and fact.

---

442 Bickford-Smith, Nicholls and Smith (n 22) 15.3.

443 Cant (n 382) 1.

7

Torts – Common Law Nuisance

7.1 Introduction

Nuisance is a:[444]

> *... tort which arises from an indirect interference with the use or enjoyment of neighbouring land. For example a nuisance may be committed where construction operations interfere with neighbouring land through the production of excessive noise, fumes, smells or vibrations.*

The PWA 1996 provides authorisation for works to party walls, which may amount to a nuisance. This would suggest that the PWA 1996 has the effect of exempting such works from the law of nuisance. However, it is not that simple.

Section 6(10) of the PWA 1996 states that:

> *Nothing in this section shall relieve the building owner from any liability to which he would otherwise be subject for injury to any adjoining owner or any adjoining occupier by reason of work executed by him.*

This would mean that:

> *The liabilities referred to here must include liability in nuisance. Thus, the building owner, it seems remains technically exposed to the possibility of liability in nuisance in respect of work carried out under s 6. But the implication is that work carried out under the enabling sections (1 and 2) are protected against liability in nuisance. More persuasive, perhaps, is the point that s 7 expressly prohibits unnecessary inconvenience and makes wide provision for loss and damage caused by any works under the Act to be compensated ... The criterion of unnecessary inconvenience under s 7(1), although apparently stricter than the nuisance test of unreasonable interference with*

444 Chynoweth (n 276) 468, 469.

Party Wall Disputes: The Party Wall etc. Act 1996 and Beyond – *Legal Coherence and Dispute Management*, First Edition. Dr Laura Lintott.

the enjoyment of land, probably approximates to it in practice ..., and the compensation recoverable for loss or damage under s 7(2) has been held to be the same as the damages available at common law. In these circumstances it is considered that for works under ss 1 and 2 liability in nuisance is technically excluded, and that for work under s 6, although not technically excluded, there is no practical scope for such liability.[445]

It has to be pointed out that:

a) The term 'unnecessary inconvenience' is not a synonym of 'nuisance' and could potentially have a wider meaning.[446]
b) The criterion of 'unnecessary inconvenience' under section 7(1) of the PWA 1996 is said to 'probably' approximate to the nuisance test of unreasonable interference with the enjoyment of land.

In any case, the common law (i.e. here the law of tort) continues to exist, but in certain circumstances defined by the PWA 1996 it is overridden by the provisions of the PWA 1996. The PWA 1996 does not codify existing common law rights; it does not repeal common law rights. It simply overrides those rights when the PWA 1996 is engaged and the relevant common law right would enable an adjoining owner to prevent the building owner's work.

In *Chaturachinda v Fairholme*:[447]

6. The Party Wall etc. Act 1996 ('the 1996 Act') makes possible the commission of what would otherwise be a trespass where the interests of adjoining landowners may conflict. It permits one landowner, the building owner, to build at the line of the junction between his land and his neighbour's, the adjoining owner, section 1. It gives a building owner a package of rights to carry out works which affect a party wall or other party structure and which at common law the adjoining owner could prevent by injunction, section 2. It also protects an adjoining owner against risks inherent in excavation and construction works carried out by a building owner in close proximity to the adjoining owner's property, section 6.

7. To achieve these ends the 1996 Act imposes a statutory scheme which regulates the work a building owner may carry out, thus affording a measure of protection to the adjoining owner, and provides a dispute resolution procedure where agreement as to proposed works cannot be reached between the two owners. The dispute resolution procedure involves the engagement of either one surveyor agreed by both owners, or in the absence of such agreement three surveyors (one appointed by each owner and a third selected by the owners' surveyors). An agreed surveyor, or all three or any two of the appointed or selected surveyors may make an award which is conclusive and binding on the two owners.

---

445 Bickford-Smith, Nicholls and Smith (n 22) 15.15.
446 *Bridgland v Earlsmead Estates Ltd* (n 101).
447 (n 80), 6–9.

> *8. The jurisdiction of the surveyor or surveyors making an award is wide. An award may determine the right to execute work, the time and manner of its execution and any other matter arising out of or incidental to the dispute, including costs and compensation. Against any award there is a right of appeal to the county court which must be exercised speedily, within 14 days of receipt of the award. On such an appeal the court is given wide powers to rescind or modify the award.*
>
> *9. Accordingly, provided he complies with the statutory scheme, a building owner has conferred on him rights which he does not have in common law. Those rights are restricted by the provisions of s.7.*

One could argue that whether it is correct to say that the PWA 1996 provides its own complete code in place of the common law, or simply overrides individual pieces of the common law is a matter for debate.

Below are some examples of case law on the interpretation of the interplay between the common law and the PWA 1996 (or its predecessors).

In *Standard Bank of British South America v Stokes*[448] it was held:

> *The building owner shall have the following rights in relation to party structures, that is to say –does not that mean he is to have no other? Is not that the definition of the rights he is to have, meaning those are all the rights he is to have? In my opinion that is the meaning of the section. The law of the metro- polis now defines the rights of a building owner.*

The implication here is that the effect of the predecessor of the PWA 1996 (in this case the Metropolitan Building Act 1855, with specific reference to section 83) was to replace the common law.

In *Selby v Whitbread & Co*,[449] McCardie J. stated:

> *The two sets of rights, namely, the rights at common law and the rights under the Act of 1894 (which followed the Act of 1855), are quite inconsistent with one another. The plaintiffs' common law rights are subject to the defendants' statutory rights. A new set of respective obligations has been introduced. The common law was seen to be insufficient for the adjustment of modern complex conditions. Hence I think that the Act of 1894 is not an addition to but in substitution for the common law with respect to matters which fall within the Act. It is a governing and exhaustive code, and the common law is by implication repealed. I follow the views on this point of Jessel M.R. in the case of Standard Bank of British South America v. Stokes (1), Warrington J. in Lewis v. Charing Cross Ry. Co. (2), and also the dictum of Collins M.R. in Leadbetter v. Marylebone Corporation. (3) I therefore hold that the plaintiffs cannot succeed upon their claim at common law inasmuch as the defendants' party wall notice had been duly given under the provisions of the Act of 1894. In so holding I in no way negative*

448 (n 5) 73.
449 (n 6), 752–753.

> *the proposition that a plaintiff may bring his action for damages if he can establish that the defendant has exerted his statutory privileges so as to inflict injury on the plaintiff by negligence, improper obstructiveness, avoidable nuisance, or unreasonable delay: see Pratt v. Hillman (4); Dodd v. Holme (5); Bower v. Peate (6); Percival v. Hughes. (7) Many authorities on this point are collected in Browne and Allan on Compensation, 2nd ed., pp. 117–119.*

The relevant legislation, and predecessor of the PWA 1996, in this case was the London Building Act 1894.

In the Court of Appeal decision in *Louis v Sadiq*[450] Lord Justice Evans stated:

> *The adjoining owner's common law rights are supplanted when the statute is invoked, which can have the effect of safeguarding the building owner from common law liabilities when he complies with the statutory procedures, just as he may incur liabilities under the statute which did not exist at common law (the Standard Bank decision). But if he commits an actionable nuisance without giving notice and without obtaining consent, he cannot rely upon a statutory defence under procedures with which ex hypothesi he has failed to comply.*

The conclusion in the above case was that the PWA 1996 supplants the common law but only if and after its mechanisms have been invoked.

In *Kaye v Lawrence*,[451] Ramsey J. (sitting as a county court judge) stated:

> *I therefore consider that in accordance with the authorities cited above, the common law rights are supplanted and substituted by the provisions of the Act and that this applies as much to sections 6(1) and 6(2) as it does to section 6(3) or any other provision of the 1996 Act. I do not consider that there is a proper distinction between rights that only exist under the 1996 Act and rights that exist at common law and which are regulated by the 1996 Act. The Act creates new rights which do not exist at common law permitting the building owner to carry out work on the land of the adjoining owner. However, whilst at common law the building owner would have a right to carry out work on their own land, those rights in relation to the area within 3 or 6 metres of the boundary are supplanted or substituted by the provisions of sections 6(1) and 6(2) of the 1996 Act.*

Again, the conclusion is that the rights under the PWA 1996 supplant or substitute the relevant common law.

The building owner is expected to comply with the PWA 1996. Should the building owner not comply (whether inadvertently or with intention), it will be in breach of statutory duty.

---

450 (n 163) 326.
451 (n 61) 61.

In *Seeff v Ho*,[452] the Court of Appeal considered an appeal in a case where no party wall notices were issued. Lord Justice Thomas stated:

> *The judge observed in his first judgment on costs that the whole dispute that followed would never have arisen if Mr and Mrs Ho had discharged their duty of giving notice under the Party Wall Act. The Act makes it mandatory to give notice in respect of work defined in the Act.*

The PWA 1996 gives the building owner rights and the adjoining owner protections not available under the common law.

As Lord Justice Coulson noted in *Power v Shah*:[453]

> *A building owner may rely on his common law rights, but in that event he is vulnerable to claims of trespass or nuisance.*

He also stated that:[454]

> *In his original skeleton argument Mr Kyson laid stress on the proposition that rights under the Act supersede common law rights. So they do, but only once the Act has been brought into operation.*

As noted earlier in this book, while a building owner's failure to trigger the notice procedure under the PWA 1996 where it is obliged to trigger it does not result in criminal liability, the building owner will be in breach of statutory duty and exposes itself to the possibility that the adjoining owner may start court proceedings against such a building owner, which can result in increased cost, time and emotion, than if solving the dispute out of court (under the PWA 1996 or via an ADR avenue).

In addition, while a party wall award can authorise works on a party wall, the building owner must still obtain the relevant consents or permissions required by any other relevant legislation. Also, an award cannot be used as an excuse for using materials that are not allowed by the relevant building regulations. One has to bear in mind that the PWA 1996 does not prevent surveyors from issuing awards allowing works that in addition need statutory consents, planning permissions or listed building consents under the Town and Country Planning Act 1990 and the Planning (Listed Buildings and Conservation Areas) Act 1990. The PWA 1996 also has no impact on building regulations requirements or work on construction sites regulations.[455] While other statutes in addition to the PWA 1996 can apply to party walls, and this is discussed in Chapter V, it is important to mention here that the rules of statutory nuisance (for example, under the Environmental Health Act 1990) continue to apply irrespective of a party wall award permitting such works under the PWA 1996.

---

452 (n 102), 36.
453 (n 2) 96.
454 ibid., 102.
455 Bickford-Smith, Nicholls and Smith (n 22) 15.26.

Undertaking building works is not, in and of itself, a nuisance, although it may be very disruptive to an adjoining owner. Only when the building owner fails to take reasonable steps to avoid unnecessary inconvenience to the adjoining owner does this constitute a nuisance.[456] If the building owner has followed the notice procedure under the PWA 1996 and complied with the relevant award, it can argue reliance upon a statutory defence. However, courts could still order for the works to stop and/or for the relevant product (e.g. party wall) to be torn down having the discretion to apportion costs as they see fit.

Works performed in connection with party walls can lead to nuisance, which is why a separate chapter on common law nuisance is included in this book. It remains to be seen what the courts' approach to nuisance and any application of the PWA 1996 will be. This chapter also considers situations where tort claims arise in connection with party walls (for example, in relation to how they are used), but not necessarily in connection with works to party walls.

7.2 The Meaning of Common Law Nuisance

Common law nuisance encompasses private nuisance (which in practice is encountered more often) and public nuisance. In *Attorney-General v Sheffield Gas Consumers Co,*[457] it was stated that:

> *... in cases of private nuisance the injury is to individual property, and in cases of public nuisance the injury is to the property of mankind.*

Private and public nuisance are not mutually exclusive. A sum of private nuisances can amount to public nuisance. The same conduct can give rise to both private nuisance as well as public nuisance.

In *Hunter v Canary Wharf Ltd,*[458] three kinds of private nuisances were identified by the House of Lords: a) nuisance by encroachment on a neighbour's land; b) nuisance by direct physical injury to a neighbour's land; and c) nuisance by interference with a neighbour's quiet enjoyment of their land. However, these are only examples of nuisance and do not constitute formal categorisation.

The cause behind private nuisance lies usually in a party's activity on its own land, such activity being lawful and which the party is entitled to do. However, such activity can turn into a nuisance when its consequences affect the neighbouring land. This can be, for example, by causing physical damage. Damage as such is not a required component of the cause of action for nuisance and is quite 'elastic' (as the Court of Appeal noted in *Network Rail Infrastructure Ltd v Williams*).[459] The damage or interference with the enjoyment of the neighbour's land a) must be substantial or unreasonable; b) can arise from a single incident or a 'state of affairs'; and c) can be caused by inaction or omission as well as by positive activity.

---

456 *Andreae v Selfridge & Co Ltd* [1938] Ch. 1.
457 (1853) 43 E.R. 119, 125.
458 [1997] A.C. 655.
459 [2018] EWCA Civ 1514.

In *Cambridge Water Co Ltd v Eastern Counties Leather Plc*,[460] it was stated:

> *But it by no means follows that the defendant should be held liable for damage of a type which he could not reasonably foresee; and the development of the law of negligence in the past 60 years points strongly towards a requirement that such foreseeability should be a prerequisite of liability in damages for nuisance, as it is of liability in negligence.*

A private nuisance can involve a continuing or repeated harm and is actionable in tort. The claimant can start civil proceedings against a defendant for either or both:[461]

a) Damages compensating for their loss for the items set out below.
   i) Actual physical damage (costs incurred or that need to be incurred to remedy the land and consequential losses).
   ii) Loss of enjoyment of the claimant's property (or proprietary interest) due to unreasonable interference with the property rights (damages awarded calculated on the basis of diminution of those rights' value as a result of the nuisance).
b) Injunctive relief requiring the defendant to abate a continuing nuisance and to ensure that its recurrence is prevented. Any compensation is for the interference with the claimant's property rights.

In *Leakey v National Trust for Places of Historic Interest or Natural Beauty*,[462] it was held that:

> *... an occupier of land owed a general duty of care to a neighbouring occupier in relation to a hazard occurring on his land, whether such a hazard was natural or man-made; that the duty was to take such steps as were reasonable in all the circumstances to prevent or minimise the risk of injury or damage to the neighbour or his property of which the occupier knew or ought to have known; that the circumstances included his knowledge of the hazard, the extent of the risk, the practicability of preventing or minimising the foreseeable injury or damage, the time available for doing so, the probable cost of the work involved and the relative financial and other resources, taken on a broad basis, of the parties; and that, in the present case, it being accepted by the defendants that the quantity and cost of the work required had not gone beyond their financial or other capacities or been greater than had been necessary to deal with the actual damage to the plaintiffs, the plaintiffs had been entitled to judgment ...*

According to the decision in *Goldman v Hargrave*,[463] the standard of the required duty of care is that the occupier is required what is reasonable to expect of it in its individual circumstances.

---

460 [1994] 2 A.C. 264, 300.
461 'Common law nuisance' (Maintained) *Practical Law Environment*, Resource ID 6-502-4804 3, https://uk.practicallaw.thomsonreuters.com/Document/Ib5554812e83211e398db8b09b4f043e0/View/FullText.html?transitionType=SearchItem&contextData=(sc.Search)#co_anchor_a698923.
462 (n 338) 486.
463 [1967] 1 A.C. 645.

7.3 Noise, Including Control of Construction Noise, Sound Proofing and Vibration

Common law noise nuisance can be private or public nuisance. In private common law noise nuisance, the claimant can start proceedings in court against the party creating the noise nuisance causing an interference with the claimant's use or enjoyment of their land. The claim is for damages and/or an injunction to stop the noise nuisance. Where an act endangers health, property or the comfort of the public, public nuisance arises, which is actionable in tort and can even amount to a criminal offence. Under Part III of the EPA 1990, a local authority will investigate a noise amounting to statutory noise nuisance. This is usually after a member of the public complains. The local authority then decides whether the issue amounts to nuisance or negatively affects health. If it decides it is a statutory nuisance, it must serve an abatement notice, which requires the recipient to stop the noise nuisance. If the recipient fails to do so, the local authority can then bring criminal proceedings against it and/or obtain an injunction to stop the nuisance. When assessing whether there is a noise nuisance, the parties should seek advice from a qualified acoustic engineer.[464]

There is no clear definition of noise nuisance, and there is also no noise level that is defined as 'prejudicial' to health. Negative effects of noise nuisance include sleeplessness and interference with personal comfort or enjoyment. The levels of noise amounting to such nuisance differ due to variability of tolerance threshold in each person. The factors that need to be judged (usually by a local authority environmental health officer for the purposes of an abatement notice) include location, time of occurrence, duration and frequency. Where a claim is started for common law private noise nuisance, it is the civil court that assesses the factors in each individual case to decide as to whether the noise amounts to nuisance or not.

In the Supreme Court decision in *Coventry (t/a RDC Promotions) v Lawrence*,[465] it was set out what approach to adopt when a court is faced with a choice between awarding damages or granting an injunction in nuisance:

> *The issues raised on this appeal are as follows: the extent, if any, to which it is open to a defendant to contend that he has established a prescriptive right to commit what would otherwise be a nuisance by means of noise; the extent, if any, to which a defendant to a nuisance claim can rely on the fact that the claimant "came to the nuisance"; the extent, if any, to which it is open to a defendant to a nuisance claim to invoke the actual use of his premises, complained of by the claimant, when assessing the character of the locality; the extent, if any, to which the grant of planning permission for a particular use can affect the question of whether that use is a nuisance or any other use in the locality can be taken into account when considering the character of the locality; the approach to be adopted by a court when deciding whether to grant an injunction to restrain a nuisance being committed, or whether to award damages instead, and the relevance of planning permission to that issue.*[466]

---

464 S Radcliffe, 'Measuring noise nuisance' Resource 0-502-1969 (2010) 1, https://uk.practicallaw.thomsonreuters.com/0-502-1969?originationContext=document&transitionType=DocumentItem&contextData=%28sc.Default%29&comp=pluk.
465 [2014] UKSC 13.
466 ibid. 6.

It was held that it was possible to obtain by prescription a right to commit what would otherwise be a nuisance by noise; a defendant could not argue that the claimant came to the nuisance; a defendant could rely on activities as forming part of the character of the locality but only as long as they did not amount to nuisance; just because the activity amounting to nuisance benefited from a planning permission would not absolve the defendant; the court has a discretion as to whether to award damages in lieu of an injunction. All these factors were relevant when the court was deciding whether to grant an injunction or damages in relation to a nuisance claim.

7.3.1 The Interrelationship Between the Party Wall etc. Act 1996 and Noise, Including Control of Construction Noise, Sound Proofing and Vibration

Paragraph 40 of DCLG's Explanatory Booklet[467] notes that a party affected by excessive noise from work being carried out in relation to a party wall should contact their local authority environmental department. This is because the EPA 1990 and the COPA 1974 give the authority the power to deal with noise and other potential nuisance issues including dust and deposits from construction sites.

When a party wall is being built, repaired, maintained or extended, disputes may arise between neighbouring owners due to the noise and vibration accompanying such construction works on the wall. Party walls also need to meet certain criteria in relation to sound insulation so that neighbouring parties are not disrupted by noise nuisance behind the party walls.

According to section 7(1) of the PWA 1996:

> *A building owner shall not exercise any right conferred on him by this Act in such a manner or at such time as to cause unnecessary inconvenience to any adjoining owner or to any adjoining occupier.*

According to section 4(3) of the PWA 1996:

> *A building owner on whom a counter notice has been served shall comply with the requirements of the counter notice unless the execution of the works required by the counter notice would – (a) be injurious to him; (b) cause unnecessary inconvenience to him; or (c) cause unnecessary delay in the execution of the works pursuant to the party structure notice.*

To clarify, the PWA 1996 authorises works that can result in physical infringement of the neighbouring owner's land or create disruptions such as dust, vibration, noise or other issues. However, the PWA 1996 also highlights the importance that this work must be carried out in such a way as to not cause unnecessary inconvenience. Such inconvenience, if it occurs, will be considered unlawful. The adjoining owners or Adjoining Occupiers then have the right to make a claim in relation to such disruption. Surveyors have a lead role in regulating the process correctly and balancing the interests of the parties by whom they

467 (n 26) 40.

have been appointed. Most importantly, it is the surveyors' duty not to frustrate the statutory process put in place. At the same time, surveyors need to ensure that the building owner can exercise its rights under the PWA 1996 and yet protect the adjoining owner's or Adjoining Occupier's interest so that no undue inconvenience is caused to them. This is done by the surveyor issuing an award, where it defines as accurately as possible what work may be lawfully carried out under the PWA 1996, including a provision requiring works to be limited to certain times and/or by imposing maximum noise/vibration limits and/or by requiring specific steps to be taken to mitigate noise, dust and/or vibration. This would then be ideally followed up by reasonable inspection of the work covered by the award.

7.3.2 Noise Nuisance Caused by Works to Party Walls

Construction and engineering noise is regulated by sections 60–61 of the COPA 1974. A local authority is responsible for serving a notice on premises imposing requirements as to how construction works are to be performed. This can include restrictions on the times of the day when these can be carried out, what plant can or cannot be used, and also defining the limits on related noise and the duration of the works. Where a party would like to prevent a local authority serving a notice, it can make a consent application predetermining the details of the works and what measures are to be applied to mitigate any issues.[468]

Transparency, communication and keeping the relevant authorities informed is a good way to mitigate the likelihood of complaints and resulting disputes. It also provides a potential defence where a statutory nuisance prosecution is started nevertheless by a local authority for breach of an abatement notice. The developer can then defend its case by highlighting that it complied with the necessary requirements. Unfortunately, this would not be considered a defence to a complaint lodged by a party, other than a local authority.[469] Typically, the local authority simply includes standard noise control measures as conditions into the panning permission, which also usually makes explicit reference to COPA 1974.

Statutory noise nuisance needs to be proven, and it is for a local authority environmental health officer to assess whether a statutory noise nuisance is being caused. It can carry out scientific assessments and compare them to the limitations in British standards. However, it is ultimately up to the environmental health officer to conclude whether they believe that a statutory nuisance is being caused. If the noise nuisance is not resolved, the local authority can serve an abatement notice, which the receiving party can appeal in the magistrates' court. If the recipient does not comply with the notice, the local authority can prosecute the recipient in the magistrates' court. Both the recipient and the local authority may need an expert witness. It is important to consider having a strong legal team as well as an experienced expert witness in order to present the best argument as to why the relevant party's expert witness' interpretation of the noise level is the correct interpretation. After hearing the arguments of each party, the magistrates' court will then conclude whether a statutory noise nuisance has been caused or not.[470]

When a party receives an abatement notice, it should check with the local authority what information it holds on the case. It should also check with its acoustic consultant whether there is a serious issue and make a full noise assessment. It is advisable to try to communicate with the local authority together with the party's acoustic consultant and

468 Control of Pollution Act 1974, s 61.
469 Environmental Protection Act 1990, s 82; Control of Pollution Act 1974, s 61(9).
470 Radcliffe (n 464) 5.

settle the matter out of court if this is at all possible. This can be done by agreeing terms acceptable to the local authority such as preparing a noise mitigation strategy that the local authority approves and then implementing it. It is advisable to keep open lines of communication and transparency between the parties to a noise nuisance dispute.[471]

7.3.3 Noise Nuisance and Party Wall Sound Proofing

Up to this point, the focus was on noise nuisance caused by works on party walls. However, party walls also have to meet certain criteria in relation to sound proofing so that neighbouring parties are not disrupted by the noise caused behind the separating wall (or otherwise called party wall). A study has been carried out in relation to the likelihood of disputes arising due to noise nuisance.

> *A random sample of the Great Britain population resident in attached properties built subsequent to 1947 was surveyed to obtain data on the incidence of nuisance occasioned by noise from neighbours, transmitted through party walls and floors. It is concluded on the basis of the present evidence that less than 10% of the population are likely to be bothered by neighbours' noise. It is nevertheless the most widespread form of noise nuisance and while the percentage of the population affected may appear relatively small the actual numbers concerned are large enough to warrant further study. There appear to be differences in the proportion of residents disturbed according to the type of dwelling, though there is little evidence of marked changes in the incidence of nuisance according to the date of construction.*[472]

The importance of preventing and/or addressing disputes arising out of noise nuisance between attached neighbouring residential properties is relevant here. Poor sound proofing between adjoining properties that leads to noise nuisance is not remediable under the EPA 1990.[473] This is because it is not reasonable to expect that neighbours behave more quietly due to the poor sound proofing between the properties. Everyday noise will therefore not amount to statutory nuisance. However, if everyday noise is created in a way that is not reasonable, such as playing musical instruments loudly for extended periods of time, this is something that can be addressed.

The Secretary of State issued guidance for the purposes of giving practical advice as to the requirements of Schedule 1 to, and Regulation 7 of, the Building Regulations 2010[474] (**Building Regulations**) for England and Wales (**Approved Document E**).[475] The aim is to achieve a basic standard of sound insulation performance internally between dwellings. This regulation provides a level of sound privacy from neighbouring parties.

Approved Document E contains specific acoustic performance requirements for new builds and conversions, relating to airborne sound insulation and impact sound insulation. The fact that refurbishments are harder to soundproof than new builds is reflected in the

471 ibid. 4 and 5.
472 F Langdon and I Buller, 'Party wall insulation and noise from neighbours' (1977) 55(4) *Journal of Sound and Vibration* 495–507. https://www.sciencedirect.com/science/article/abs/pii/S0022460X77811747?via%3Dihub.
473 Following case law, in particular *Southwark LBC v Mills* [2001] 1 A.C. 1, *Baxter v Camden LBC* (No. 2) [2001] Q.B. 1; *R. (on the application of Vella) v Lambeth LBC* [2006] Env. L.R. 33.
474 SI 2010/2214.
475 HM Government, 'Resistance to the passage of sound – Approved Document E' (2003 edn incorporating 2004, 2010, 2013 and 2015 amendments).

requirements, which are more lenient towards refurbishments. The requirements relate to sound insulation that is between residential dwellings sharing a party wall or floor. They do not relate to internal floors and walls inside dwellings. These fall under more lenient requirements for sound insulation that does not have to be sound tested. Sound insulation relates to airborne and impact sound insulation. Airborne sound insulation relates to sound travelling through the air and through separating structures between individual rooms. Sound that this relates to includes speech, music and television sound, i.e. noise sources radiating sound into the air and not directly into the structure. Impact sound insulation relates to noise caused by an actual impact on a structure, for example, footsteps, chair scrapes, dropped objects or objects impacting on the floor.

In order for residential dwellings to comply with Building Regulations, a qualified acoustic engineer[476] must conduct acoustic testing. Diagrams 7.1–7.3 set out below show sets of sound tests required for different types of residential dwellings or developments. They relate to types of testing required for flats with a separating floor and party wall, flats with separating floors and dwelling-houses with party walls but no separating floors (the electronic version of this book contains coloured images for Diagrams 7.1, 7.2, and 7.3).

Sound testing should be conducted once building works have been completed. Usually, where an acoustic test has failed, this is due to the construction works not having been completed or where penetrations and air gaps have not been sealed and doors not put in place. Any impact sound insulation testing in residential dwellings should be done before the floor finish to ensure this is accurate as different types of floor finish can influence the test results, which may no longer apply where floor finish is replaced with a different material (for example carpets and hard flooring).[477]

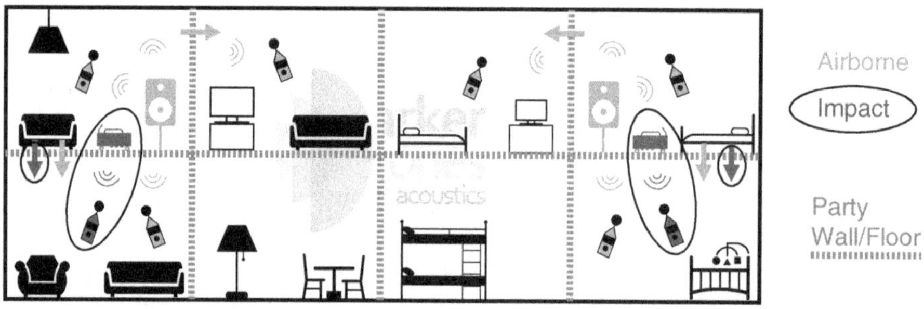

One set of tests should compromise 6 tests (4 airborne, 2 impact):

- 1 airborne test across a wall between (where possible) a pair of living rooms
- 1 airborne test across a wall between (where possible) a pair of bedrooms
- 1 airborne and 1 impact test across a floor between (where possible) a pair of living rooms
- 1 airborne and 1 impact test across a floor between (where possible) a pair of bedrooms

Diagram 7.1 Sets of tests in flats with a separating floor and party wall.[478]

476 Accredited to conduct sound testing by an organisation such as UKAS or the Association of Noise Consultants.
477 C Parker-Jones, 'Sound Testing to comply with Part E for Residential Developments', https://www.parkerjonesacoustics.com/insights/articles/sound-testing-explained?_sm_au_=isVqLCN6nnV6w05VQcLJjKQ1j7GJ1. Please see coloured images in relation to Diagrams 7.1, 7.2 and 7.3 on the aforementioned website.
478 ibid.

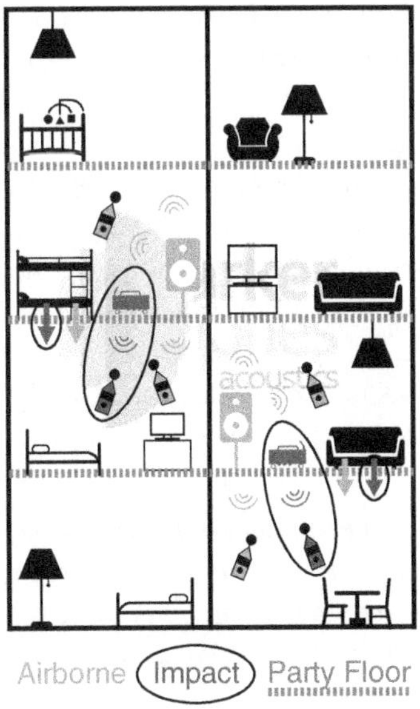

One set of tests should compromise 4 tests (2 airborne, 2 impact):

- 1 airborne and 1 impact test across a floor between (where possible) a pair of living rooms
- 1 airborne and 1 impact test across a floor between (where possible) a pair of bedrooms

Diagram 7.2 Sets of tests in flats with separating floors.[479]

One set of tests should compromise 2 tests (2 airborne):

- 1 airborne test across a wall between (where possible) a pair of living rooms
- 1 airborne test across a wall between (where possible) a pair of bedrooms

Diagram 7.3 Sets of tests in dwelling-houses with party walls but no separating floors.[480]

7.3.4 Noise Nuisance and Party Walls – Case Law Examples

7.3.4.1 'Reasonable' Use of Dwellings

The case law examples set out below show the importance of distinguishing between what is considered to be reasonable and unreasonable noise in the context of party walls and why.

The first case here is that of *Reinhardt v Mentasti*.[481] Although this case is about a stove put up by the defendants, the heat of which penetrated through the party wall and rendered

479 ibid.
480 ibid.
481 (1889) 42 Ch. D. 685.

the cellar of the adjoining house unfit for storing wine, Kekewich, J. commented on the issue of noise explaining the relevance of its degree when deciding whether it amounts to nuisance.

As for the case, Kekewich, J. held that, although the defendants acted reasonably in the way in which they used their house, they did in fact cause serious annoyance and injury to the claimant. The claimant claimed for an injunction and damages. Kekewich, J. granted an injunction; however, this was not enforceable for three months as Kekewich, J. gave the defendants time to consider what is the best course of action for them in the given circumstances.

As for nuisance and the relevant test in relation to the level of disturbance by one party to the other when it starts to amount to nuisance, Kekewich, J. noted:

> *It seems to me, therefore, that, notwithstanding some passages in some judgments to the contrary, the application of the principle governing the jurisdiction of the Court in cases of nuisance does not depend on the question whether the defendant is using his own reasonably or otherwise. The real question is, does he injure his neighbour?*[482]

This statement shows that while one party's action may be reasonable for its own use, that does not prevent the action from amounting to nuisance if it injures the other party. Kekewich, J. refers to another case in his decision, namely *Broder v Saillard*,[483] where he reiterates a quotation from Jessel, M.R. set out below.[484]

> *I take it the law is this, that a man is entitled to the comfortable enjoyment of his dwelling-house. If his neighbour makes such a noise as to interfere with the ordinary use and enjoyment of his dwelling-house, so as to cause serious annoyance and disturbance, the occupier of the dwelling-house is entitled to be protected from it.*

Kekewich, J. further refers to the same case and Jessel, M.R.'s quotation as to the issue that the 'reasonable use of its property' cannot be the test where the party materially disturbs the other party from enjoying its property.

> *... whether the stables are unluckily so situated as that the noise from the horses, not being uncommon horses in any way, materially disturbs the comfort of the plaintiffs' dwelling house, and prevents the people sleeping at night.*[485]

It is clear that, while a party may not be unreasonable in the level of noise it is creating on its premises, the crux lies in the fact whether the noise is such as to 'materially disturb' the adjoining owner from also enjoying its property. It is a question of degree and circumstances in each case.

---

482 ibid. 690.
483 (1876) 2 Ch. D. 692.
484 ibid. 689.
485 ibid.

In *Southwark LBC v Mills*,[486] which was a House of Lords decision, the focus was on the relevance of adequacy of insulation in relation to party walls in residential dwellings. In the first case, the tenants made a complaint under the arbitration clause of their tenancy agreement that the sound insulation between the flats was not effective as it did not minimise the noise created by the normal and ordinary user of the premises. The arbitration tribunal held that the landlord (council) was in breach of the covenant for quiet enjoyment in each tenancy agreement and ordered the landlord to carry out the necessary works to the premises to rectify the position. The council appealed to the High Court and the judge upheld the arbitration tribunal's award. The council appealed again to the Court of Appeal, which by a majority reversed the judge's decision.

In the second case, the tenants complained to the council that the noise created by the neighbours' day-to-day living seriously interfered with the claiming tenant's enjoyment of the flat due to inadequate sound insulation. What is meant by 'day-to-day living' of the neighbours is the reasonable use of domestic premises. The council refused to improve the sound insulation. As a result, the tenants brought an action in the county court claiming damages for breach of the covenant of quiet enjoyment of the tenancy agreement and for nuisance. The court dismissed the claim and the Court of Appeal upheld the county court's decision.

The tenants proceeded to appeal to the House of Lords, which dismissed the appeal on the grounds set out below.

a) The covenant for quiet enjoyment did not extend to interference due to the condition of the property before the grant of the tenancy or from usage that the parties must have contemplated. The noise that the tenants complained about was due to, what essentially amounted to structural defects at the time the flats were let to the tenants. The noise also resulted from activities within the tenants' contemplation. Therefore, the landlord was not in breach of the covenant for quiet enjoyment.
b) The ordinary and reasonable use of the flats by the neighbours did not amount to an actionable nuisance. Where a tenant complained of noise nuisance emanating from the neighbour's flats, the common landlord could not be held liable in tort as a result of having permitted the commission of an actionable nuisance. This is the case where the noise complained of stems from ordinary and reasonable use where the neighbour could not be held liable for such noise nuisance.

Focussing on the issue of noise nuisance in this chapter, there are a number of points of relevance when establishing what can amount to noise nuisance and why. The noise complained of in question included the neighbour's televisions, babies crying, coming and going, cooking, cleaning or quarrels. The resulting lack of privacy caused tension and distress to the tenants.

One of the statute-related points noted by Lord Hoffmann relevant to this case included the fact that Mrs Tanner's block of flats was constructed in about 1919. Ms Baxter's terrace house was converted in 1975. This meant that both sets of works (construction and conversion) had to comply with the byelaws in force at the time (i.e. the London Building Acts). The issue there is that these were silent on the required standards regarding sound insulation.

486 (n 473).

The Building Act 1984 updated and replaced the previous system of local byelaws by introducing nationally applicable regulations prepared by the Secretary of State for the Environment. Specifically, the Building Regulations 1985 (SI 1985/1065) included for the very first time a requirement that walls and floors separating dwellings from each other should resist the transmission of airborne and impact sound.[487] However, the reason why these regulations could not be applied by the House of Lords to the case is that the regulations apply only to buildings constructed or converted after the regulations came into force. This means that, as the relevant buildings in the case pre-date the regulations, the tenants could not rely on them. This is also the reason why the tenants attempted to rely on the argument of noise nuisance. When Lord Hoffmann turned to the issue of the law of private nuisance, he noted that the main crux of the problem the tenants were dealing with was the fact that:[488]

> *Nuisance involves doing something on adjoining or nearby land which constitutes an unreasonable interference with the utility of the plaintiffs' land.... What is the nuisance of which the appellant's complain? The sound emanating from their neighbours' flats. But they do not allege the making of these sounds to be a nuisance committed by the other tenants.... But I do not think that the normal use of a residential flat can possibly be a nuisance of the neighbours. If it were, we would have the absurd position that each, behaving normally and reasonably, was a nuisance to the other... . I do not understand how the fact that the appellants' neighbours are living in their flats can in itself be said to be unreasonable. If it is, the same, as I have said, must be true of the appellants themselves.*

7.3.4.2 Noise Caused by Construction Works

In *Matania v National Provincial Bank Ltd*,[489] the second defendant carried out alterations to its first-floor premises without obtaining the consent of the first defendant, who was their landlord. The alterations resulted in works creating excessive dust and noise. The claimant had to vacate its premises, which were on the first and second floors, respectively, of the property as shown in Diagram 7.4 below.

The Court of Appeal held that the first defendant (landlord) was not in breach of its covenant for quiet enjoyment. This is because the first defendant did not carry out the works and did not permit them either. It was the second defendant's contractor who failed to exercise reasonable skill and care when carrying out the works.

As for the dust, the Court of Appeal held that this could not have been avoided. It also held that the noise could have been mitigated had the contractor staggered working hours avoiding most disruptive times for the claimant. The contractor's conduct resulted in actionable nuisance and the second defendant owed a duty that could not be delegated to the contractor[490] in relation to the works, considering these had an inherent risk of nuisance. The second defendant was therefore liable for its contractor's negligence.

---

487 Part E of Schedule 1 to the Regulations.
488 *Southwark LBC v Tanner* [2001] 1 A.C. 1, 15.
489 [1936] 2 All E.R. 633.
490 *Jolliffe v Woodhouse* (1894) 10 TLR 553.

This case shows that noise caused by works that carries through the party wall can result in nuisance, the liability of which does not necessarily fall on the party causing it directly (here the contractor). Works to a party wall carry with them an inherent risk of damage to adjoining property, which gives rise to a non-delegable duty of care. Otherwise, liability would lie solely with the contractor.

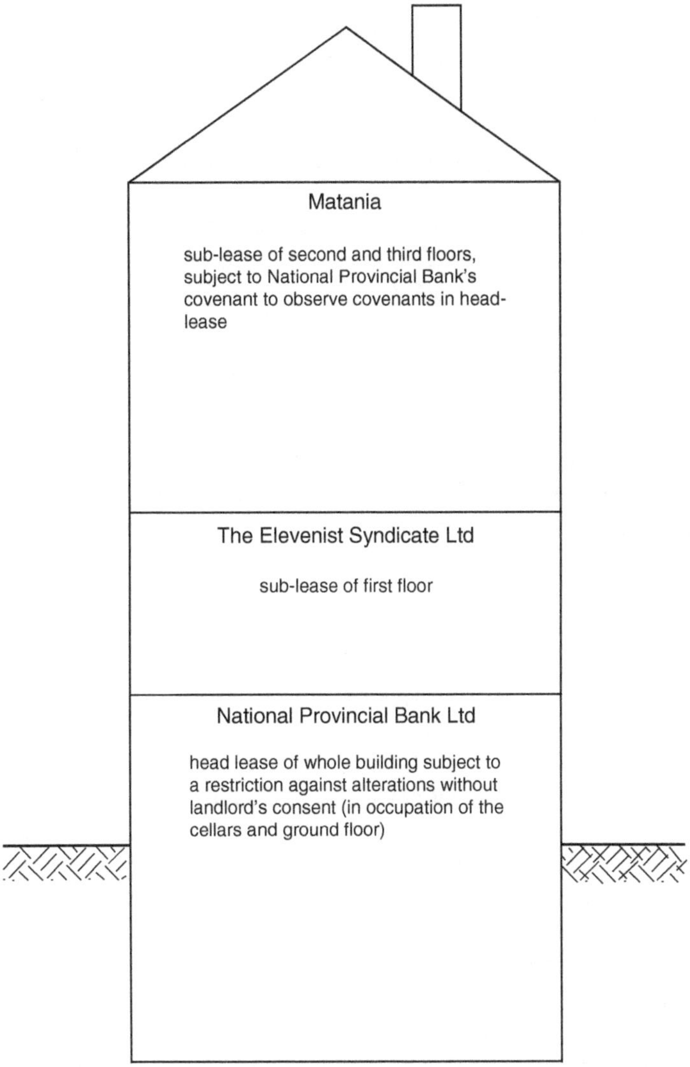

Diagram 7.4 *Matania v National Provincial Bank Ltd.*[491]

In *Hough v Annear*[492] the parties owned adjoining properties and one party wanted to repair the roof of its property to build an extension. The parties instructed a joint surveyor

491 Chynoweth (n 276) 277.
492 [2007] 11 WLUK 31.

who concluded that the PWA 1996 applied to the works in question and later added that the works however did not affect the party wall. The parties therefore agreed for the works to progress. Once the works started, however, the claimant started complaining about the noise and dust emanating as a result of the works. It also turned out that the works in fact did impinge on the party wall. Once the works were completed a surveyor prepared a party wall award and a determination. According to the claimant, the works involved an intrusion and therefore the defendant was in breach of section 3 of the PWA 1996 resulting in the claimant's entitlement to damages. The damages related to cracking to the claimant's property, noise nuisance and dust. The noise complained of related to the works allegedly having been carried out too noisily and the radio playing too loudly.

The court held that damages did not apply for breach of statutory duty under section 3 of the PWA 1996 as there was no significant impact on the value of the claimant's property. However, it was accepted that cracking did occur during carrying out of the construction works, and the claimant was therefore awarded £600. As for the noise, it was held that the works carried out prior to 6 p.m. were not considered to be too noisy. There was no evidence of the presence of a radio. It was accepted that there was some excessive noise on occasions after 6 p.m. and before 9 p.m. It was rejected that there was an unreasonable amount of dust. As a result, the claimant was awarded £350 in relation to the issue of nuisance. Finally, it was also held that there was no substantive claim in trespass.

This case suggests that the issue of what is considered to be 'excessive' noise and what is considered to be 'reasonable' noise lies at the crux of whether the court will consider the noise in relation to a party wall to amount to nuisance and award any damages and/or an injunction.

7.3.5 Vibration

Similar to the restrictions on noise and dust, the PWA 1996 authorises work involving physical encroachment onto a neighbour's land or which produces vibration (or other inconvenience or annoyance) as long as the work is carried out in a way that does not amount to unnecessary inconvenience. If work has been authorised by an award, then any unnecessary inconvenience caused can be the subject of a claim for compensation under section 7(2) of the PWA 1996. It is advisable for the adjoining parties to attempt to discuss the matter first before proceeding to formal dispute resolution. Alternatively, the parties can consider negotiation through their consultant surveyors (this is different from appointing a surveyor for the purposes of a determination under the PWA 1996).

Vibration caused by works to party walls can cause, for example, the brickwork and mortar to shake. This in turn can cause loosening of materials such as dust or debris. Sometimes bricks can be loosened and enter the property of the adjoining owner. This can result in dirt on the adjoining owner's property leading to costly cleaning bills.

The PWA 1996 makes sure that, should the building owner's party wall works cause damage to an adjoining owner's party wall, the building owner will be held responsible under the PWA 1996 for the cost of making good such damage. It is clear that, prior to starting party wall works, the building owner should consider what proactive steps it can take so that its contractor follows these in compliance with the PWA 1996. Protection requirements are invariably contained in a substantive award before any damage has occurred.

For example, the use of hand tools only is one of the standard requirements or conditions often attached to an award authorising works. Party wall surveyors will impose such conditions in the award without reference to the building owner directly.

Section 79(1)(a)–(h) of the EPA 1990 includes categories of matters, which can amount to a statutory nuisance. These include noise from premises, as well as vibration emitted from premises so as to be prejudicial to health or be a nuisance.

In *Scott v Firth*,[493] it was stated that:[494]

> *In order to constitute a nuisance, there must be, not merely a nominal, but such a sensible and real damage as a sensible person, if subjected to it, would find injurious, regard being had to the situation and mode of occupation of the property injured. Semble, if a man builds a rolling mill close to inhabited cottages, so that the vibration produced by the hammers cracks the walls of the cottages, and the noise of the mill causes them to become and remain uninhabited, he cannot justify these injuries, on the ground that they were caused by him in the reasonable and proper exercise of his trade in a reasonable and proper place.*

This case was referred to in *Reinhardt v Mentasti*[495] (noted above in connection with noise nuisance) and gives an example, albeit not relating to a party wall, where vibration is severe enough to constitute nuisance. The facts of the case related to an action for a nuisance. The claimant complained that its several cottages and reversion suffered an injury (the walls of the cottages cracked, the tenants left and the cottages remained unoccupied) caused by vibration stemming from the steel hammers used in the workshops of the defendant. The claimant also complained of the disturbance caused by the noise of the hammers. The defendant argued that the grievances the claimant complained of were caused by the defendant in the reasonable and proper exercise of his trade in a reasonable and proper place. In addition, the defendant argued that the cracking of the cottage walls was caused by the removal of the adjoining cottages (taken down by the defendant to be able to build the mill) and not by the vibration. In addition, the defendant argued that the unoccupied cottages were dilapidated and in disrepair. Therefore, they were unoccupied due to their state and not the noise. Blackburn, J. stated:[496]

> *The question is, whether this is a case of nuisance, that is, of actionable wrong? If the defendant, in the course of using these hammers, produced, not merely a nominal, but such a sensible and real damage as a sensible person occupying the cottage would find injurious, that is, a nuisance; but that which is a sensible and real inconvenience to property situated in one place, or occupied in one way, will be none to property situated in another place or occupied in another way. If you are of the opinion that the vibration caused by the hammers has shaken and cracked the walls of the cottages, you will*

493 176 E.R. 595, 595.

494 ibid. 595.

495 (n 481); *Hiscox Syndicates Ltd v Pinnacle Ltd* [2008] EWHC 145 (Ch) involved notifiable demolition works where the vibrations were severe enough to justify an injunction.

496 *Scott v Firth* (n 493) 596.

> *probably consider that to be a substantial and real mischief. If, on the other hand, you think the damage was caused by the removal of the adjoining cottages, whether that was justifiable or not, you ought to find a verdict for the defendant on that part of the case. So with regard to the cottages standing empty: if that was caused by the hammering, you will find a verdict for the plaintiff; if by the want of repair, for the defendant. A further point has been raised by the plea that the grievances complained of were caused by the defendant in the reasonable and proper exercise of his trade in a reasonable and proper place. My opinion is, that in law that is no answer to the action. I think that that cannot be a reasonable and proper exercise of a trade which has caused such injury to the plaintiff as she complains of. That, however, is matter of law to be disposed of hereafter; and the defendant is entitled to your opinion upon the fact, whether this was a reasonable and proper exercise of the trade, although it did cause injury to the plaintiff. I am unable to direct you as to what is reasonable and proper in that sense. You must say, first, was there a nuisance to the plaintiff's property, being such as it is? secondly, if you are of opinion there was a nuisance, what is the amount of the damages? thirdly, were those damages caused by the reasonable and proper exercise of the defendant's trade.*

The jury found for the defendant based on the first point; therefore, it became unnecessary to consider the rest of the issues. However, the points raised by Blackburn, J. assist in assessing when and why vibration may amount to actionable nuisance. One factor that needs to be considered is the degree of real damage that a sensible person would consider harmful. This needs to be decided on a case-by-case basis, as what constitutes a real inconvenience amounting to nuisance in one situation may not amount to nuisance in another. A further factor is whether the grievance has been caused "*in the reasonable and proper exercise of a party's trade in a reasonable and proper place.*"[497] Blackburn, J. concluded that this is not relevant. The point is that even if the grievance has been caused "*in the reasonable and proper exercise of a party's trade in a reasonable and proper place,*" that does not justify the injury.

While not relating to party walls, in *Hiscox Syndicates Ltd v Pinnacle Ltd*,[498] the balance of convenience favoured a claimant who sought an injunction to prevent excess vibration during the defendant's demolition works. This is because damages would not be an adequate remedy, and without the injunction the claimant would have no protection from the nuisance.

In *Video London Sound Studios Ltd v Asticus (GMS) Ltd and Keltbray Demolition Ltd*,[499] the second defendant was a demolition contractor who was instructed by the first defendant to demolish buildings on its site. The demolition works resulted in substantial damage to some expensive recording equipment of the claimant - see Diagram 7.5. The damage was partly caused by vibration to the building due to the demolition works. The surveyors that were appointed issued an addendum award, which required the first defendant to reimburse the claimant the cost of the damage. However, the first defendant did not pay. Therefore, the claimant brought proceedings in nuisance, negligence and for breach of the surveyors' award. The Technology and Construction Court held that the surveyors'

---

497 ibid. 595.
498 (n 495).
499 [2001] 3 WLUK 133.

addendum award was in fact invalid. This is because the recording equipment was not a fixture but a chattel. Therefore, the first defendant was not held to be liable for breach of the award and was also not liable in negligence for failing to undertake an adequate pre-demolition inspection. The court however held that both defendants were liable in nuisance as the defendant's demolition works caused physical damage to the claimant's property. The recording equipment did not form part of the premises of the claimant as it was not annexed to the premises and a chattel could be treated as part of realty only if some degree of physical annexation was present. A reasonable care and skill defence did not apply considering the claimant suffered physical damage. The damage was reasonably foreseeable.[500]

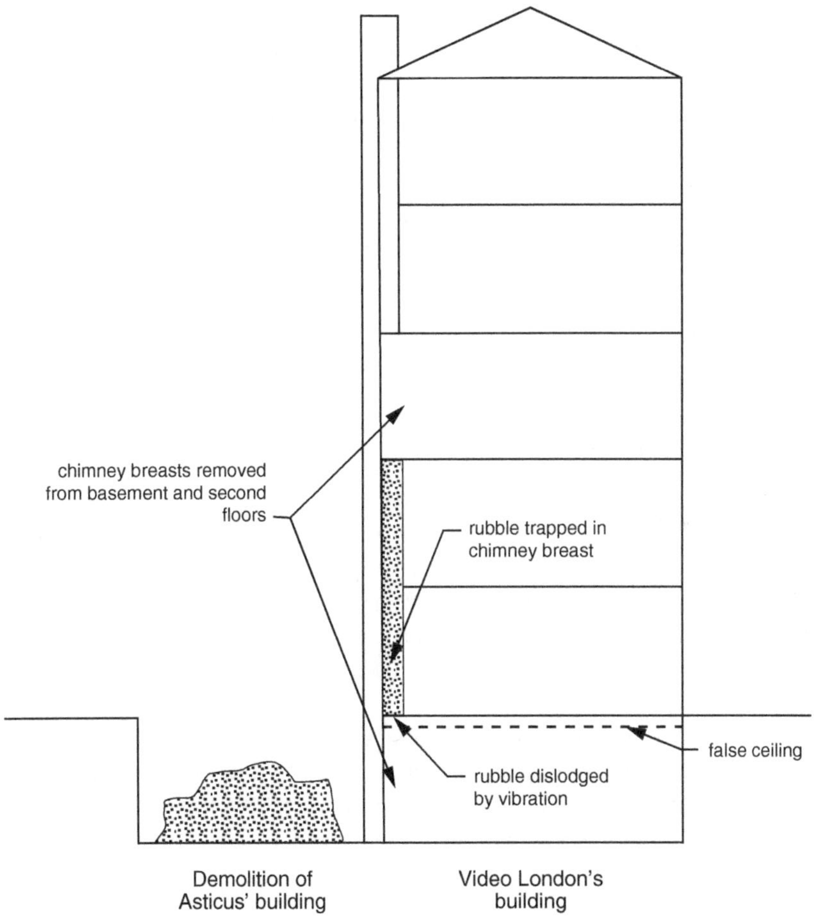

Diagram 7.5 *Video London Sound Studios Ltd v Asticus (GMS) Ltd and Keltbray Demolition Ltd.*[501]

500 Chynoweth (n 276) 406.

501 Chynoweth (n 276) 408.

7.3.6 Noise, Including Control of Construction Noise, Sound Proofing and Vibration – Conclusion

As can be seen above, noise and vibration can come hand in hand in connection with works to party walls covered by the PWA 1996. Use of a party's property can also cause noise and vibration. One of the key factors in deciding whether the noise or vibration amount to nuisance is whether these are substantial enough.

It is not a sufficient defence that the defendant used its own property reasonably. The deciding point is whether the defendant's behaviour injures its neighbour.[502] The law is that a party is entitled to the comfortable enjoyment of its dwelling. It is therefore protected from noise and vibration caused by its neighbour interfering with its ordinary use and enjoyment of its dwelling causing serious annoyance and disturbance.[503]

It follows that a party's reasonable use of its property must not disturb its neighbour from enjoying its property.[504] If the noise or vibration materially disturbs the adjoining owner from enjoining its own property, it will amount to nuisance irrespective of whether such disturbance was caused by reasonable use of the party's property causing the noise or vibration.

However, each case needs to be assessed individually as the level and severity of noise/vibration amounting to nuisance is not defined and is subjective to the facts or circumstances of every individual case. Courts will look at and assess the degree and circumstances in each case.

7.4 Rights to Light

7.4.1 What is a Right to Light?

A right to light is an easement where the benefiting party has the right to enjoy the natural light, which passes over another party's land and then enters through defined openings (such as windows, glass roofs or skylights) into a building. As confirmed in *Colls v Home & Colonial Stores Ltd*,[505] once such a right to light has been established, it entitles the benefiting party to receive enough natural light through such opening as to allow the room or space behind the opening to be used for its ordinary purpose. The level of a right to light is limited in scope.

Land that has not been developed cannot obtain or benefit from a right to light across the land of another. This means that when it comes to land that has not been built on, there is no right to light. In accordance with the decision in *Dalton v Henry Angus & Co*,[506] there has to be a building on the dominant land benefiting from having access to light. The right to light refers to light providing natural illumination and not to direct sunlight.

---

502 *Reinhardt v Mentasti* (n 481).
503 *Broder v Saillard* (n 483) 701.
504 ibid. 702.
505 [1904] A.C. 179.
506 (n 340).

There is no right to a 'prospect or view' according to the decision in *Phipps v Pears*.[507]

> *Suppose you have a fine view from your house. You have enjoyed the view for many years. It adds greatly to the value of your house. But if your neighbour chooses to despoil it, by building up and blocking it, you have no redress. There is no such right known to the law as a right to a prospect or view ...*

There is no right to a view to a property:[508]

> *The erection of a building will not be restrained because it injures the Plaintiff by obstructing the view of his place of business.*

A right to light does not protect privacy:[509]

> *It appears to me that to constitute a breach of such a covenant there must be some physical interference with the enjoyment of the demised premises, and that a mere interference with the comfort of persons using the demised premises by the creating of a personal annoyance such as might arise from noise, invasion of privacy, or otherwise is not enough.*

A right to light does not entitle a claimant to the same amount of light that they enjoyed before an obstruction was built. There can be a reduction of light subject to what is left being enough to enjoy the ordinary and normal use of the relevant room.

As for solar panels, a right to light does not apply to direct sun rays to fuel them. This is because a panel situated on the roof does not constitute a room used normally. This approach has been adopted by the Law Commission:[510]

> *... objects that do not have apertures, such as solar panels, are almost certainly not capable of benefiting from a right to light.*

The Law Commission commented:[511]

> *Accordingly, we do not make any recommendation that allows the creation of easements that benefit solar panels.*

One of the issues is that solar panels have currently unclear protection from future neighbour development. Considering that solar panels have been now used for a while, one

507 (n 351) 83.
508 *Butt v Imperial Gas Co* (1866-67) L.R. 2 Ch. App. 158, 158.
509 *Browne v Flower* [1911] 1 Ch. 219, 228.
510 Law Commission, 'Reforming the law – Rights to Light' 356, 24, https://assets.publishing.service.gov.uk/government/uploads/system/uploads/attachment_data/file/391684/44872_HC_796_Law_Commission_356_PRINT.pdf.
511 ibid.

could pose the question whether these could acquire a right to light by prescription. This is something that will eventually be decided by courts.

As noted earlier, right to light is an easement where one party has the right to enjoy the natural light passing over another party's land entering through defined openings. The right to light is limited to dominant land that has been built upon and to natural illumination (not direct sunlight). A reduction of light is considered to be acceptable as long as enough light is left for the ordinary enjoyment of the relevant room. The easement of a right to light can be created in the immediate term by implied grant or by express grant by way of a deed (or reservation) or by statute. It can be created also by reason of use from *time immemorial*. In practice, rights of light are acquired under the Prescription Act 1832 (**PA 1832**) or under the doctrine of lost modern grant. Enjoyment of a right to light can be interrupted in the process of being acquired by prescription. A right to light that has already been acquired can be expressly or impliedly abandoned.

7.4.2 Nuisance Related to the Right to Light

Although the right to light is an easement, obstruction of passage of light can constitute a nuisance in tort. In *Higgins v Betts*,[512] the court considered what the test was for whether an obstruction to light amounts to actionable nuisance:

> ... *the test is not whether so much light has been taken as materially to lessen the enjoyment and use of the house that its owner previously had, but whether so much is left as is enough for the comfortable use and enjoyment of the house according to the ordinary requirements of mankind.*

Where it is found that the right to light was or will be substantially interfered with, the party whose land is the owner of the right to light has the option to apply for an injunction against the infringing developer. It should be taken into account that the court has discretion to award damages instead of awarding an injunction. In *Beaumont Business Centres Ltd v Florala Properties Ltd*,[513] while the injunction was considered to be appropriate, the judge determined the compensation which he would have awarded if an injunction had not been appropriate. In general, a court may be inclined to award damages instead of an injunction where it deems an injunction to be inappropriate.

7.4.3 Enforcing a Right to Light

Rights to light can be easements and the breach of these can result in nuisance. In common law nuisance, the claimant has to prove that the defendant caused the nuisance and that it damaged the claimant's land or substantially interfered with its property interest. To be able to prove nuisance stemming from interference with a right of light, the affected party must show that the loss of light amounts to a nuisance, and that the interference

512 [1905] 2 Ch. 210, 210.
513 [2020] EWHC 550 (Ch).

was substantial and of a level entitling the owner of the right to light to equitable remedy (such as injunction or specific performance) and/or compensation. Different measurement methods of light adequacy are discussed further below. It is not enough to show that the amount of available light has been reduced. This has been confirmed in *Higgins v Betts*:[514]

> *There must be a substantial deprivation of light, sufficient to make the premises in a sensible degree less fit for the purposes of business or occupation.*

Where a right to light has been or is about to be infringed, the party benefiting from the right to light has the right to apply for an injunction against the other party causing such interference. The court can however grant damages instead of an injunction. Where a claimant is to establish a nuisance, it has to prove that, due to the reduction in light, its property has become substantially less comfortable and convenient than before the reduction in light has occurred. This test has been established in *Beaumont Business Centres Ltd v Florala Properties Ltd*:[515]

> *Accordingly, I conclude that to establish its claim in nuisance, Beaumont needs to prove that, by virtue of the reduction in light, its premises have been made substantially less comfortable and convenient than before; which in practice means it must show that, by virtue of the reduction, it is likely to suffer a loss of rental income over the balance of its 26 year term, in an amount which is more than trifling or de minimis. It is therefore irrelevant whether the premises were "well lit," within the definitions used by rights of light surveyors, before the reduction in light caused by Florala's works.*

Although there are no standard measures of quantifying what amounts to 'sufficient light', there are 'rules of thumb' used as a point of reference. The Building Research Establishment (**BRE**) guidelines (**BRE Guide**)[516] aim to ensure that adequate levels of natural light can be achieved in new dwellings, and unacceptable impact on light to neighbouring properties is minimised. Local Planning Authorities, which use BRE, only usually approve planning applications which do not have a negative impact on daylight and sunlight to neighbouring properties. The BRE Guide provides advice on site layout planning to achieve good sunlight and daylight, both within buildings and in the open space between them. The intention is that the BRE Guide is to be used together with the British Standard Daylight in buildings, BS EN 17037. This covers guidance on site layout to ensure good natural lighting within a new development, safeguarding of daylight and sunlight within existing buildings nearby and the protection of daylighting of adjoining land for future development.

---

514 (n 512) 212.
515 (n 513) 138.
516 Building Research Establishment, 'Site layout planning for daylight and sunlight: A guide to good practice' (2022), https://bregroup.com/store/bookshop/site-layout-planning-for-daylight-and-sunlight-a-guide-to-good-practice-br-209-2022-edition.

> *C1 The British Standard "Daylight in buildings" (BS EN 17037) contains advice and guidance on interior daylighting. The guidance contained in this publication (BR 209) is intended to be used with BS EN 17037 and its UK National Annex[C1]. Other European countries have their own versions of EN17037, which do not include the UK National Annex.*
>
> *C2 BS EN 17037 supersedes BS 8206 Part 2 "Code of practice for daylighting," which contained a method of assessment based on Average Daylight Factor, which is now no longer recommended. For daylight provision in buildings, BS EN 17037 provides two methodologies. One is based on target illuminances from daylight to be achieved over specified fractions of the reference plane (a plane at table top height covering the room) for at least half of the daylight hours in a typical year. The other, alternative, method is based on calculating the daylight factors achieved over specified fractions of the reference plane.*
>
> *C3 BS EN 17037 gives three levels of recommendation for daylight provision in interior spaces: minimum, medium and high. For compliance with the standard, a daylit space should achieve the minimum level of recommendation.*[517]

A separate section of the BRE Guide addresses loss of solar radiation for solar panels and for passive solar buildings using the sun as a source of heating energy. Guidance is also provided on the sunlighting of gardens and amenity areas. Issues like privacy, enclosure, microclimate, road layout and security are briefly covered. The appendices also explain methods to quantify access to sunlight and daylight within a layout.

According to the BRE Guide:[518]

> *The amount of daylight a room needs depends on what it is being used for. But roughly speaking, if θ is:*
>
> *– greater than 65° (obstruction angle less than 25° or VSC at least 27%) conventional window design will usually give reasonable results.*
>
> *– between 45° and 65° (obstruction angle between 25° and 45°, VSC between 15% and 27%) special measures (larger windows, changes to room layout) are usually needed to provide adequate daylight.*
>
> *– between 25° and 45° (obstruction angle between 45° and 65°, VSC between 5% and 15%) it is very difficult to provide adequate daylight unless very large windows are used.*
>
> *– less than 25° (obstruction angle greater than 65°, VSC less than 5%) it is often impossible to achieve reasonable daylight, even if the whole window wall is glazed.*

There is no 'hard and fast rule' in relation to the angle of 45° as a matter of law.[519] The reality is that a person can be in a situation where it is left with much less than 45° of light, but still has suffered no actionable diminution.[520]

---

517 ibid. 72.

518 ibid. 9.

519 *Colls v Home and Colonial Stores Ltd* (n 505).

520 *Gale on the Law of Easements* (n 11) 7–27.

Another 'rule of thumb' is the '50:50 rule':

> *A method employed by expert witnesses to measure the obscuration of light caused by a new building is to estimate the amount of direct sky which will reach a hypothetical table 2 feet 9 inches high in a particular room. By this method a room is regarded as adequately lit for all ordinary purposes if 50 per cent or more of its area receives not less than one lumen of light at table level. This so-called "50/50 rule," however, is merely a useful guide and not to be applied rigidly without regard to the shape and size of the room or the disposition of the light within it. Its justification is that an owner is unreasonable if he complains that the corners or other parts of the room where good light is not expected are poorly lit, if the room as a whole remains well lit. The proportion of the room which receives the required amount of light is determined by the drawing by a rights of light surveyor of a light contour map known as a "Waldram" diagram after Mr Percy Waldram, who devised the method in the 1920s.*[521]

While a special form of Waldram diagram can be used for estimation on an external wall or window as an alternative to the skylight indicator,[522] the Waldram diagram approach has been criticised in *Sheffield Masonic Hall Co Ltd v Sheffield Corp*[523] and *Fishenden v Higgs and Hill*[524] as not being conclusive and in *Ough v King*[525] it was held that the method was not decisive.

It is important to keep in mind the risks of an injunction being granted when the building owner ignores complaints from the adjoining owner, something that the BRE Guide highlights:[526]

> *As a general guide, currently if after construction of a proposed development more than half of a room in an existing building has a sky factor of less than 0.2%, and a right to light exists, then this right will probably be infringed. Developers should exercise extreme caution as the law in this area is very complex and they should seek expert advice. The risks of an injunction being granted are considerable, particularly if complaints from adjoining owners are ignored.*

7.4.4 The Interrelationship Between the Party Wall etc. Act 1996 and Rights to Light

The test of an interference is not how much light was taken away by an obstruction but whether there is enough light left for the ordinary enjoyment of the dominant tenement.

---

521 ibid. 7–28.
522 Building Research Establishment (n 516) 9 and 68.
523 [1932] 2 Ch. 17.
524 (1935) 153 L.T. 128.
525 [1967] 1 W.L.R. 1547.
526 Building Research Establishment (n 516) 4.

The amount of light to which a building is entitled is that which is enough for the ordinary use of the building.[527]

> ... *the owner or occupier of the dominant tenement is entitled to the uninterrupted access through his ancient windows of a quantity of light, the measure of which is what is required for the ordinary purposes of inhabitancy or business of the tenement according to the ordinary notions of mankind, and that the question for what purpose he has thought fit to use that light, or the mode in which he finds it convenient to arrange the internal structure of his tenement, does not affect the question.*[528]

Section 9 of the PWA 1996 sets out its relationship with rights to light below.
Nothing in this Act shall-

- *(a) authorise any interference with an easement of light or other easements in or relating to a party wall; or*
- *(b) prejudicially affect any right of any person to preserve or restore any right or other thing in or connected with a party wall in case of the party wall being pulled down or rebuilt.*

Although section 9 of the PWA 1996 aims to cover easements in general, it also does specifically refer to and protects the rights of light. It follows that, although section 2(2)(l) of the PWA 1996 gives the right "*to raise a party fence wall, or to raise such a wall for use as a party wall, and to demolish a party fence wall and rebuild it as a party fence wall or as a party wall,*" this cannot interfere with an existing easement of light.[529]

Although the PWA 1996 is clear on the need to protect the easement of light, when surveyors are faced with a dispute and having to make an award, they may come across pertinent legal questions before they are able to make such an award. It needs to be established whether there is an easement at all. In addition, if there was an easement once, the point may be raised that the easement has come to an end, for example by way of abandonment, and to prove abandonment there has to be evidence of an intention to abandon.[530] However, in practical terms, it is unlikely for a party wall surveyor to be willing to conclude that a right to light had been abandoned by an adjoining owner.

7.4.5 Case Law Examples of Party Walls' Ability to Interfere with Rights to Light

To probe the connection between party walls and rights to light, below are examples of relevant case law showing how party walls may interfere with rights to light and what stance courts might take in cases concerning or touching upon party walls. In summary, the examples show situations related specifically to parties' sincerity and conduct, evidence of parties' agreements, court's discretion to choose between awarding damages and injunctive relief, whether a greenhouse can be considered to be a building and therefore benefits

527 Bickford-Smith, Nicholls and Smith (n 22) 15.7–15.8.
528 *Colls v Home & Colonial Stores Ltd* (n 505) 204.
529 *Crofts v Haldane* (n 389); Bickford-Smith, Nicholls and Smith (n 22) 15.8.
530 Bickford-Smith, Nicholls and Smith (n 22) 15.8.

from rights to light or not, whether a wall can be divided into a party wall and an 'ordinary wall' containing windows, whether a party wall can be subsequently extended in terms of height and how this may interfere with the neighbour's right to light.

In *Ottercroft Ltd v Scandia Care Ltd*,[531] the parties occupied adjoining properties. Without having served a notice on the claimant under the PWA 1996, the defendants started construction works on a storeroom and a staircase. The claimant considered the defendants' works interfered with the claimant's right to light. Therefore, the claimant started court proceedings. This was followed by each of the defendants giving an undertaking. However, the defendants continued and completed building a metal fire-escape staircase. This was in the way of the claimant's view. The judge at first instance held that the staircase obstructed the claimant's window and therefore infringed the claimant's right to light. The defendants' actions were in breach of the undertaking and were without notice under the PWA 1996 or a planning permission. Although the infringement was considered to be minor so that the damage could have been measured in money, the judge at first instance held that the breach of binding undertakings gave a reason to grant an injunction. He ordered that the staircase be removed or altered and awarded costs to the claimant. He also held that the second defendant kept the claimant in the dark regarding his plans despite knowing that these might interfere with a right to light. The defendants appealed the first instance decision on the grounds that the order was disproportionate and oppressive[532] noting that damages should have been awarded rather than an injunction, that the defendants were not personally liable and that the judge should not have ordered payment of the claimant's action, party wall legal and surveyor's costs. The Court of Appeal dismissed the appeal and decided to grant a mandatory injunction instead of damages due to the inconsiderate conduct as well as the fact that the staircase was installed in breach of the defendants' undertakings.

> *... the judge considered that the defendants had acted badly throughout, and earlier in his judgment he had described Dr Rahimian's behaviour as "high handed"; indeed, he went so far as to hold that Dr Rahimian wished his neighbours to remain in ignorance of what he was doing.*[533]

On the point of proportionality, when considering remedies, these are measured against what is appropriate to the particular situation. In this case, damages might have been appropriate if the defendants' conduct had been conscionable. However, the defendants acted in a high-handed and oppressive manner, and in such a situation they could not have expected for the court to exercise its discretion to their advantage. It should be noted that, where developers are trying to negotiate with neighbour(s) and the neighbour(s) refuse(s) to participate in the negotiation, it is advisable that developers maintain an evidence trail

---

531 [2016] EWCA Civ 867.

532 Procedure: obligations to the court' (2016), 37(6) *Property Law Bulletin*, 44, https://uk.westlaw.com/Document/IED200F109B7C11E6BC4BD8CA38210D45/View/FullText.html?transitionType=SearchItem&contextData=(sc.Search)&firstPage=true.

533 *Ottercroft Limited v Scandia Care Ltd* (n 531) 13.

showing their offers and suggested alternatives, as well as any lack of cooperation or failure to respond on the part of the neighbour(s). Developers should also maintain evidence of their advance notice of proposed works. Such evidence is more likely to win the court's sympathy.[534]

In *Frederick Betts Ltd v Pickfords Ltd*,[535] the claimant leased a plot of land from the defendant. This land was partially covered with buildings and occupied by the defendant together with the adjoining property as business premises. The lease had a covenant obliging the claimant to build a warehouse. This would have to be done based on approved plans showing that the back wall of the warehouse had to include certain windows. The wall of the warehouse was built entirely on the leased land. The local authority then asked the claimant to block up windows in the back wall of the warehouse on the basis that the wall was a party wall.

The court held that the claimant had an implied unqualified right to the access of light passing through the relevant windows and that the claimant's architect who supervised the building of the warehouse lacked the authority to block the windows. The defendant used the wall as a party wall causing the local authority to demand that the windows had to be blocked up. It was further held that the defendant derogated from its grant by using the wall as a party wall and trespassed. As a result, the court ordered that the defendant disconnects its buildings from the wall.

The first floor of the warehouse was to be used as a showroom with three windows in the south or back wall, which were to overlook the defendant's premises. There were no planned windows for the back in the ground floor of the warehouse. After some initial objections, the defendant eventually approved the building plans. Under a collateral agreement, the defendant was obliged to clear buildings off the leased land.

When the back wall of the claimant's warehouse was built, the claimant's architect verbally agreed with the defendant's architect, without the claimant's knowledge, for the projecting roof beams and stanchions to be built into the wall, which was done. The claimant was later served with a notice by the district surveyor under the London Building Act 1894 requiring the claimant to brick up all three windows in the back wall as otherwise the warehouse would fail to comply with the Act.

The claimant was summoned for not having complied with the notice. At the summons hearing, the magistrate opined that the back wall of the warehouse (to the extent covered by the defendant's shed or extended within three feet above the roof of the shed) was a party wall. Therefore, the claimant would have to block up the relevant windows. However, the summons was adjourned pending trial. The claimant's main argument was that blocking up the relevant windows would lead to the showroom being unfit for its intended use.[536]

> *The plaintiffs claimed: (1.) An injunction to restrain the defendants from committing any trespass on the walls of the warehouse and from using these walls as party walls; (2.) a mandatory injunction requiring the defendants to disconnect their buildings*

534 Procedure: obligations to the court' (n 532).
535 [1906] 2 Ch. 87.
536 ibid. 89–90.

from the plaintiffs' walls; (3.) an injunction to restrain the defendants from preventing the plaintiffs from enjoying the lights of the back of the warehouse according to the terms of the lease, and a mandatory injunction requiring the defendants to pull down any erections which interfered with such enjoyment; (4.) damages.

Reference was made to the decision in *Drury v Army and Navy Auxiliary Co-operative Supply, Ltd*[537] where it was held that a wall used as a party wall does not always lead to the wall being a party wall in its entirety. The party wall is only a party wall to the extent of the user, for example, only a part of the wall up to a certain height may be a party wall.

Kekewich J. noted that it is a matter of fact that the parties agreed for the claimants to erect buildings containing certain windows. Kekewich J. then proceeded to refer to a decision by Stirling J. in his judgment in *Aldin v Latimer Clark Muirhead & Co.*:[538]

... *where a landlord demises part of his property for carrying on a particular business, he is bound to abstain from doing anything on the remaining portion which would render the demised premises unfit for carrying on such business in the way in which it is ordinarily carried on.*

Considering that blocking off the relevant windows would lead to the claimant's showroom not being fit for its purpose, following Stirling J.'s line of thought, the defendants were wrong in saying that the wall was in its entirety a party wall.

Kekewich J. further stated that the claimant's architect agreed with the defendants' architect that the roof beams and stanchions should be built into the wall, which was also done. However, the claimant's architect did not communicate this to the claimant, who found out only after this was done. Such action was not within the claimant's architect's authority.

It was held that the external wall was built by the claimant on its own land. The defendants had no right to use the claimant's wall. In fact, the defendants used the claimant's wall as a party wall, which the defendants had no right to do. In addition, the claimant protested as soon as it found out about this.

The court's decision is set out below.[539]

There will be a perpetual injunction restraining the defendants from committing any trespass on the walls of the plaintiffs' warehouse and from using any part of the walls as a party wall, and an order that the defendants do within one month from the service thereof remove the stanchions and roof beams so far as they stand upon or overlap the plaintiffs' premises, and there will be also an inquiry as to damages. The defendants to pay the costs of the action down to and including the trial.

537 (n 15).
538 [1894] 2 Ch. 437, 444.
539 *Frederick Betts Ltd v Pickfords Ltd* (n 535) 97.

The main point of this case lies in the fact that a party cannot choose to use its neighbour's wall that is built on the neighbour's land as a party wall, especially if the neighbour protests as soon as it has knowledge of such use.

This case also touches on the question of whether a section of a wall can be simply a wall and not a party wall, and whether another section of the same wall can be a party wall. Although this was raised in this case, it was not answered as it did not form the crux of the case. However, the same question was raised in an older case: *Crofts v Haldane*,[540] where it was decided that the simple answer is yes, a part of the same wall can be a party wall while the rest of the wall is not a party wall (and is just a simple wall instead). The two sections of the wall can co-exist in harmony. The part of the wall that is not a party wall can contain windows. The case also highlights the seriousness with which courts approach a party's insincerity and that the court has discretion to choose between injunctive relief and damages. When it comes to interference of a party wall with a party's rights to light, courts will look at what the parties agreed amongst themselves, their conduct as well as the actual function of the party wall and whether it is actually a party wall and to what extent.

Considering dispute management related to a clash between party walls and rights to light, parties would be wise to keep any agreements, communications and complaints, as well as offers of settlement in writing and accessible, should the matter proceed to court for decision as courts will take the parties' conduct and recorded purpose of the disputed wall into consideration. *Frederick Betts Ltd v Pickfords Ltd*[541] is a prime example of intertwined issues between parties linked to party walls and rights to light.

Another case of particular interest is that of *Clifford v Holt*[542] as it relates to a greenhouse and whether it was considered to be a 'building' within the meaning of section 3 of the PA 1832 that is capable of enjoying a right to light and also whether it can be protected by an injunction against interference with access to light.

The claimant was the occupier of a detached house and garden under a lease. At the back of the house was the garden attached to it, which was bounded on its North West side by a party wall or fence of about eight feet, which divided it from the garden of another house purchased by the defendant. There was a greenhouse at the further end of the claimant's garden and a section of the party wall formed the northern end of the greenhouse. Both the claimant's house and the greenhouse enjoyed a right to light.

The defendant started to build a party wall in the garden without giving any notice to the claimant. The claimant was not at home at the time and had no knowledge of the defendant's intention to build the party wall. When the claimant returned and saw the party wall, he consulted his architect. The architect contacted the defendant's architects noting that the party wall was causing significant obstruction to the access of light and air to the claimant's greenhouse and that unless the building works were stopped subject to further negotiations with the claimant, he would apply for an injunction. As the defendant was not forthcoming, the claimant applied for an injunction:

> *... from erecting or permitting to remain erected any wall, building, or other structure, so as to damage, injure, or obstruct any of the lights of the plaintiff's house or*

540 (n 389).
541 (n 535).
542 [1898] 12 WLUK 27, 698.

greenhouse, and from unlawfully raising or otherwise dealing with the party wall between the plaintiff's and defendants' premises or trespassing on the said wall.[543]

The defendant raised the wall from 8 to 25 feet and intended for the building works to continue so that the party wall would reach 40 feet. The claimant alleged that this would cut off the north and northeast light to his greenhouse and would also interrupt access of air to his premises and lower their value.

Kekewich J. proceeded to assess whether the greenhouse could be considered to be a 'building' within the meaning of section 3 of the PA 1832 requiring access of light for ordinary purposes for which light is required:[544]

> *... when a building is of such a character that access of light is of importance to it, it is in fact a "building," and that therefore the Legislature intended that there should be protection for that light if it has been enjoyed for an uninterrupted period of twenty years.*

It was held:

> *... that the defendants' wall caused a substantial obscuration of light to the plaintiff's greenhouse, and that the obscuration was of that material character that the Court ought to interfere; also that the existence of the wall was sufficient to render the plaintiff's residence less marketable ... it has been established to my satisfaction that the wall does interfere with the plaintiff's greenhouse, and on that ground he is entitled to the protection of the Court.*[545]

The claimant was granted a mandatory injunction. This aligns with the view of the Law Commission and RICS, which view greenhouses to be capable of enjoying a right of light (therefore being considered a 'structure' for this purpose capable of acquiring rights to light by prescription).[546]

Yet again, it can be seen that, as long as the claimant can show that its land (under ownership or lease) has a right to light which has been interfered with by the defendant, courts tend to exercise their discretion to award an injunction to stop such interference as damages are not always enough to make good the damage caused by such an interference.

In *Presland v Bingham*,[547] the claimant initially started an action for an injunction or damages for interference with a right to light pertaining to its house as the defendant re-built an old party wall 15 feet above its original height. This was however not due to a solid structure but due to the defendant piling up empty cases against the party wall, which were from time to time removed, i.e. the obstruction was not permanently in place. As the evidence was conflicting, North, J. at first instance held that there was no

543 ibid. 699.
544 ibid. 702.
545 ibid. 702–703.
546 Law Commission, 'Reforming the law – Rights to Light' (n 510) 36, 39, 45; RICS professional guidance, UK, 'Rights of light' (2nd edn, 2016) 4.
547 [1889] 3 WLUK 16, 268.

interruption of the claimant land's right to light, and that the claimant had a right to an inquiry as to damages.

North, J. held that, first, the claimant proved that it had a right to light for more than twenty years which has been materially interfered with by the defendant's new wall. The judge then proceeded to the second question as to what the effect of the piled-up boxes was. He concluded that, while the claimant had a right to light, there was no evidence that this had been interrupted. He therefore did not grant any relief to the claimant. However, due to the judge's dissatisfaction with the defendant's evidence, he dismissed the action without costs.

The claimant appealed. Cotton, L.J. commented on the burden of proof of each party:[548]

> *I may say my own opinion is that if it appears upon the evidence of the Plaintiffs that there has been an interruption of a permanent character, one likely in its nature to be of a permanent character, then it would lie on them to shew that it did not in fact last for a year; but if on the evidence it appears that the interruption is one not likely to be of a permanent character, one which from its very nature is not of a permanent character, it lies upon the Defendant, who objects to the light having been gained, to shew by his evidence that there has been an interruption which has been existing and acquiesced in for more than a year.*

Cotton, L.J. then proceeded to analyse the nature of the 'interruption' and concluded that the interruption was only of a temporary nature and therefore the defendant's defence failed.

The conclusion was that:[549]

> *... the access of light to the skylight over the workshop of the Plaintiffs has actually been enjoyed for the full period of twenty years before the commencement of this action without any such interruption ...*

Cotton, L.J. held that any damage was to be ascertained by the Official Referee in relation to the present wall having been built above the height of the old party wall of eight feet.

This case is slightly different compared to typical party wall cases in the context of a right to light in that here the dispute revolved around a temporary infringement of less than a year. By temporary infringement, it is meant piling up boxes against the old party wall (not newly built party wall) as the boxes had not remained in the same place for a year due to the fact that they had been removed and replaced with other boxes over the course of time. It was due to the temporary nature of these piled-up boxes that damages were deemed appropriate as opposed to a mandatory injunction.

The case of *Smith v Smith*[550] concerns a landowner and occupier of a house and workshop (claimant) and his neighbour (defendant) who raised a low party wall positioned eight feet from the claimant's window to a height of 26 feet. The claimant started proceedings as it complained about the party wall creating an obscuration of its right to light. The court held

548 ibid. 274.
549 ibid. 276.
550 (1875) L.R. 20 Eq. 500.

that the claimant did not lose its right to relief by delay or agreement. Therefore, the claimant was entitled to a mandatory injunction to remove the party wall.

In this case, the claimant was able to produce evidence that the rooms facing the party wall were deprived of light to such an extent as a result of the raised party wall that the affected rooms could be used only with the assistance of gaslight and that the workshop was close to not being fit for its purpose. The evidence also showed that the house was next to uninhabitable, the health of some inmates was affected and that the claimant's family had to leave the premises. The defendant tried to argue that the claimant's right to light was barred by delay in asserting his claims and acquiescence.

Sir G. Jessel, M.R. concluded that the court has discretion to award damages in place of a mandatory injunction where it considered fit to do so. He further held that the judicial discretion was to be exercised so "*... as to prevent the Defendant doing a wrongful act, and thinking that he could pay damages for it.*"[551] One decisive factor to consider was whether the defendant was aware that he was doing wrong and was taking on the risk of being disturbed in doing the wrong. Another important (however not standalone) factor was whether the claimant was materially injured. However, this needs to be assessed in context with the amount laid out by the defendant.

Sir G. Jessel, M.R. held that the injury in this case was most serious to the claimant and that damages were not sufficient.

In this case, one can see that a party that is the owner of a land enjoying a right to light that has been infringed (and showing relevant evidence) has a good chance of pursuing an injunction against the defaulting party. One can act early so that the defaulting party does not succeed in trying to argue that the party benefiting from a right to light has lost it through delay in opposing the infringement or even by acquiescence. However, there is also a drawback in 'acting early' in the sense of seeking an interim injunction as the claimant will be required to give a cross-undertaking in damages. Consequently, the usual approach is to notify the defendant developer of the prospective claim, but to commence proceedings, if at all, for a permanent injunction only, thereby avoiding the need to give a cross-undertaking.

The decision confirms, again, that the courts have discretion to choose between awarding damages or an injunction.

Here the defendant was fully aware that it was depriving the claimant of the right to light by erecting the party wall. The claimant in this case was materially injured and it was held that the defendant could not 'buy' its infringement, i.e. doing a wrong and then simply paying damages instead of an injunction. It would have been out of proportion for the defendant to be allowed, in essence, to force the claimant to accept the defendant's infringement of the claimant's right to light and for the claimant to leave the house and be forced to 'sell' it to the defendant.

On the principle of the courts having discretion to award damages in lieu of an injunction, while this is still the case and has been further confirmed in the Supreme Court's judgment in *Coventry (t/a RDC Promotions) v Lawrence*,[552] the approach to that discretion has been subject to debate over a number of years (with some examples set out below).

551 ibid. 505.
552 (n 465).

- In *Shelfer v City of London Electric Lighting Co (No. 1),*[553] it was held that the courts' discretion to award damages in lieu should not be exercised in order to deprive a claimant of its right to an injunction (with rare exceptions).
- In *Miller v Jackson,*[554] it was held, on appeal, that when the court exercises its equitable jurisdiction to grant an injunction, it must take into consideration public interest.
- In *Coventry (t/a RDC Promotions) v Lawrence,*[555] it was held that public interest has to be taken more into account, where an existing planning permission authorising carrying on an activity causing a nuisance can be a factor favouring the court to grant damages in lieu of an injunction and that a more flexible approach should be taken by courts when being asked to award damages in lieu (as opposed to the approach suggested in *Watson and others v Croft Promo-Sport Ltd,*[556] where judges have been too keen to grant injunctions without considering the option of granting damages in lieu).
- In *Raymond v Young,*[557] the Court of Appeal pointed out that there may be circumstances where both damages at common law and a permanent injunction could be awarded (this could be the case where, for example, it is likely that the defendant's conduct will continue).

It is safe to say that party walls are very well capable of infringing a right to light hence causing disputes between parties. While going to court is an option, it is often worthwhile for parties to consider whether settling the matter out of court is a better option, thereby mitigating additional costs and the protracted amount of time linked to court proceedings, especially if the costs linked to the party wall are disproportionately small compared to costs linked to court proceedings for each of the parties.

In *Weston v Arnold*[558] the claimant was the owner of a house where one wall was to the height of the first storey a party wall between the claimant's house and the defendant's building. However, above that height, the wall had windows opening to the external air. The claimant demolished his house and wished to rebuild it with windows in the same position as prior to the demolition. He therefore gave a notice to the defendant that the wall, which the claimant viewed to be a party wall, was in disrepair and was to be rebuilt at the parties' joint expense as supported by two surveyors. The defendant then built a building obstructing the light coming through the claimant's windows as per Diagram 7.6 below. At first instance, it was held that the wall above the defendant's building was in fact not a party wall and the claimant was not prevented from making windows in it. An injunction was granted to prevent the defendant from obstructing the right to light.

553 [1895] 1 Ch. 287.
554 [1977] 3 W.L.R. 20.
555 (n 465).
556 [2009] EWCA Civ 15.
557 [2015] EWCA Civ 456.
558 (n 245).

The wall between the parties' premises is described below.[559]

> *The south side of the Plaintiff's house overlooked a courtyard and outbuildings belonging to the Defendants, which outbuildings were built up against the Plaintiff's house, so that to the height of the first story the wall of Plaintiff's house was a party wall between the two buildings. The Plaintiff's house consisted of a ground floor, used as a kitchen, and two stories above it, and there were twelve windows in the south wall, all of which opened on the external air over the outbuildings. The Defendants' building now complained of was a warehouse just beyond the outbuildings.*

The defendant was of the opinion that, when the wall was built again, it should not have any windows as it is a party wall. After the injunction was granted at first instance, the defendant requested that the Vice-Chancellor dissolve the injunction. This was refused on the basis that the part of the wall that extended above the defendants' buildings was not a party wall. The defendant then proceeded to the Court of Appeal.

Sir W. M James, L.J. held that:

> *Now, a party wall is a thing which belongs to two persons as part owners, or divides two buildings one from another. It is beyond even the power of the Legislature to make that a party wall which is not a party wall… . they cannot make what is not a party wall a party wall, any more than they can make a square a circle.*[560]

He held that in this case, part of the wall divided the two houses and the part above it was the claimant's separate and undisputed property:[561]

> *… it appears to me that a wall may be in part of its length a party wall, and in part of its length an external wall, and there is no distinction between height and length.*

As the windows were in the separate wall above the party wall section, the claimant as the owner has acquired a right to the windows in this separate wall and a neighbour could not raise his building to darken such windows. Sir W.M. James, L.J. concluded that the claimant was entitled to the injunction obtained from the Vice Chancellor. Sir G. Mellish, L.J. agreed with Sir W.M. James, L.J. and confirmed that Weston had a right to light, and that the injunction has to be made perpetual against the defendant.[562] This decision is consistent with that in *Crofts v Haldane*[563] as well as in *Frederick Betts Ltd v Pickfords Ltd*[564] discussed above.

559 ibid.1085.
560 ibid. 1089.
561 ibid. 1090.
562 ibid. 1093.
563 (n 389).
564 (n 535).

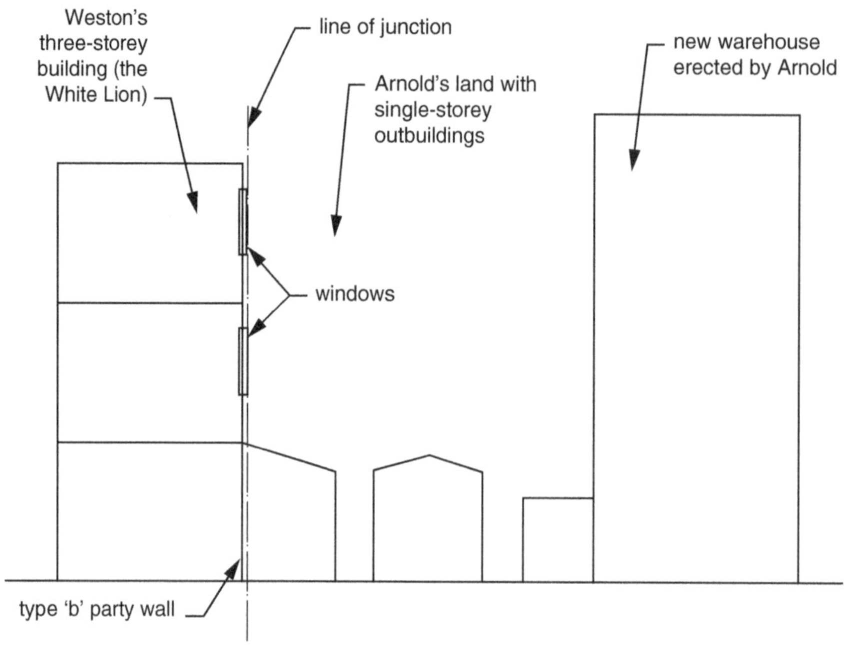

Diagram 7.6 *Weston v Arnold.*[565]

As noted in Chapter III, the oldest piece of legislation covering party walls resembling the modern form is Part III of the Metropolitan Building Act 1855.[566] The case of *Crofts v Haldane*[567] is from the year 1867 and although it relies on the Metropolitan Building Act 1855, the main principles, when it comes to party walls, remain the same under the PWA 1996. In this case, the claimant was the owner of premises used as a gallery for exhibiting pictures and paintings for sale as well as water-colour drawings and other works of art. For the gallery to be functional, it needed to have the light and air entering through certain windows. The defendants wrongfully erected a party wall and building near such windows. This prevented the claimant (adjoining owner) from using his premises as a gallery.

A dispute arose between the parties and was referred to three surveyors which produced an award stating that the party wall and building "*had been built in conformity with the award, and part of the erection was the wall and building complained of by the plaintiff.*"[568] However, the actual dispute about light, air and related damage remained unsettled and to be decided by the surveyors.

Cockburn, C.J. held that:

> *... amongst other rights in relation to party structures, the building owner shall have a right to raise any party structure permitted by the act to be raised, upon condition of*

565 Chynoweth (n 276) 420.
566 (n 49).
567 (n 389).
568 ibid. 196.

> *making good all damage occasioned thereby to the adjoining premises, or to the internal finishings and decorations thereof. Under that enactment, Mr. Clarke contends that the building owner has a right to raise an external wall, although he interferes with the ancient lights in the neighbouring premises... . the statute gives the right to raise any party structure "on condition of making good all damage to the neighbouring premises and the internal finishings," it clearly means that the building owner must restore the neighbouring premises to their condition previous to the erection; and cannot be forced into meaning, not only that he is to restore or make good any structural damage, but to give pecuniary compensation for destroying access to light ... unless express power be given to interfere with the right to light, it ought not be inferred.*[569]

Cockburn, C.J. also noted that:[570]

> *... if the building owner has no right to raise a party wall so as to interfere with the right to light, there is no matter arising under the act for the surveyor's arbitration.*

The right to build a party wall does not give a right to interfere with the right to light (section 9 of the PWA 1996). While party walls are physically capable of interfering with a right to light, they are not legally entitled to do so. It is also clear that courts have a discretion to award an injunction or damages in relation to a party's interference with another party's right to light. Courts are not hesitant to award an injunction where they conclude that damages do not suffice in a situation where a right to light exists, has not been extinguished and has been interfered with without the interfering party having the right to do so.

7.4.6 Rights to Light – Conclusion

While the right to light is an easement, its infringement can constitute a nuisance in tort. As noted in *Colls v Home & Colonial Stores Ltd*,[571] the test establishing whether a right to light has been infringed is whether enough light is left for the comfortable use and enjoyment of the relevant premises in line with the usual requirements of mankind. If the landowner of the dominant tenement has suffered an infringement of its right to light, they can apply for an injunction against the infringing party, bearing in mind that the court has the discretion to award damages instead where it considers an injunction to be inappropriate and damages to be a sufficient remedy. This has been confirmed also in *Ottercroft Ltd v Scandia Care Ltd*,[572] *Frederick Betts Ltd v Pickfords Ltd*[573] and *Presland v Bingham*.[574]

Considering the case law examples in this chapter, parties need to be aware that, should their case get to court, courts are likely to take into account the parties' conduct and any

569 ibid. 198.
570 ibid.
571 (n 505).
572 (n 531).
573 (n 535).
574 (n 547).

agreements amongst parties themselves in relation to the right to light and/or party wall in question. Courts have assessed which parts of a wall are a party wall and which are not. The question is whether there is an aperture in a structure which has been in place for a sufficient period of time to acquire prescriptive rights (if an express grant has not been made). For the purposes of rights to light, for example in *Clifford v Holt*,[575] the court held that a greenhouse in that instance did amount to a building.

Where the owner of the dominant tenement wishes to enforce a right to light, they must show that the loss of light amounts to nuisance, i.e. that there is a substantial deprivation of light making the premises less fit for their intended purpose[576] or less comfortable and convenient than before the reduction in light occurred.[577] There is no given measure quantifying what amounts to 'sufficient light'. However, there are 'rules of thumb' that can assist in assessing the required amount of light.

Section 9 of the PWA 1996 states that the PWA 1996 does not authorise any interference with an easement of light. As can be seen from case law on the interrelationship between party walls and rights to light, as long as it is established that the right to light exists, has not been extinguished, and has been interfered with, courts tend to make an award in favour of the dominant tenement usually in the form of an injunction or, if this is not appropriate, in the form of damages, at the discretion of the court.

Before going to court however, parties are advised to consider whether it is appropriate to settle disputes related to party walls infringing rights of light outside of court by way of ADR. This is because court proceedings take time and cost money, which can go over and above the value of the issues in dispute in the context of party walls. Parties, usually owners or occupiers of neighbouring properties, have to consider that they may have to co-exist with the opposing party to the dispute for years to come unless one of the parties decides to leave. Therefore, the relationship between the parties needs to be taken into consideration. The best way forward is transparency and clear communication between the parties before any issues arise. Where one party intends to raise a party wall, it is prudent to check if such a wall could potentially infringe the other party's right to light. The party planning to build such a wall should notify the other party of its intentions and consider any concerns the benefiting party raises in advance of any construction works on the relevant party wall. Both parties should keep a record showing communications between them, recording the chronology of events and how the situation develops. This will add clarity and assist any settlement negotiations as well as potential court proceedings.

However, it is not always possible to avoid a clash between neighbouring properties when it comes to party walls being built that infringe a right to light. In a situation where a property is about to acquire a right to light due to having had 20 years of uninterrupted enjoyment of light under the PA 1832, a developer has the option to serve a timely light obstruction notice before the 20 years expire as then it will prevent the right to light from being acquired. This is subject to the notice not being successfully disputed by the benefiting owner resulting in the notice being set aside.

---

575 (n 542).
576 *Higgins v Betts* (n 512).
577 *Beaumont Business Centres Ltd v Florala Properties Ltd.* (n 513) 138.

At an early stage, and prior to building a party wall that may infringe a neighbouring property's right to light, the developer should consider investigating applicable insurance covers for the risk of any claims that may potentially arise. The cost of such a cover is a factor that the developer should consider as well. The insurance cover will not lower the chances of a dispute, but it may cover full or part of the consequences of such a claim for infringement of rights to light. The exact wording of the insurance policy needs to be carefully checked as not every insurance cover covers the same issues, and the cost linked to such cover may vary considerably as well as its limits. Insurers usually require that the policy holder does not make contact with potential claimants (contrary to courts' views). Therefore, insurance covers of this type may be limited to small interferences with rights to light. Hence, where insurance does not apply, developers should try to approach the benefiting owner to try and negotiate a settlement, potentially for a monetary price. Alternatively, the developer can contact the local authority who may agree to extinguish the rights.[578] The threat of this step alone may convince the parties to settle the matter amongst themselves rather than going through official channels.[579]

On the other hand, the benefiting owner needs to issue a notice to the developer as soon as the benefiting owner becomes aware of an infringement of a right to light. As noted above, although disputes are sometimes unavoidable, parties usually benefit from settling the matter amongst themselves both from a time as well as cost perspectives. There are several ADR avenues available as discussed in Chapter IV of this book.

578 Town and Country Planning Act 1990, s 237.
579 Hogan Lovells, 'Practice & law – developing themes' (2013) *The Estates Gazette, ABI/INFORM Collection*, 120.

8

Structural Issues

Party walls are essentially structures that can affect neighbouring properties. While this book has been focussing on the legal aspects and context of party walls, it is not possible to discuss party walls without mentioning practical issues such as structural issues linked to party walls, which can result in a dispute if not managed with a dispute prevention/resolution strategy in mind. Structural issues range from cracks in party walls, subsidence, underpinning, thickening, raising, repairing, cutting into a party wall, increasing or decreasing its height, to exposing a party wall or demolishing it and fully rebuilding it.

Section 2(1) of the PWA 1996 states:

> *This section applies where lands of differing owners adjoining and at the line of junction the said lands are built on or a boundary wall, being a party fence wall or the external wall of a building, has been erected.*

Subsection 2(2) of the PWA 1996 sets out the building owner's rights. It is the works to which section 2 of the PWA 1996 applies that require a notice to be served prior to starting the works.

Cracks could mean a grave structural issue in a party wall. An element of movement in houses is inevitable potentially resulting in cracks in the plasterwork or brickwork. It can result from poorly considered alterations, which weakened the structure due to inadequate support during works or because the works have added weight or stress to the walls, which was not expected at the time the building was built. One can sometimes observe structural problems prior to serious damage or collapse of the structure, the signs of which include: cracks and bulges in the internal/external walls or binding doors/windows due to structural movement or damp weather, tapered cracks running diagonally from the corner of doors/windows, which point to weak spots in the structure, cracked render/plaster around the top of windows suggesting that the carrying part of a timber lintel is rotting, cracks between a bay window and the building suggesting incorrectly tied bay windows to the structure and sloping floors indicating subsidence or other structural issues.

Party Wall Disputes: The Party Wall etc. Act 1996 and Beyond – *Legal Coherence and Dispute Management*, First Edition. Dr Laura Lintott.

Costs of repairing or rebuilding a party wall are to be shared between the building owner and the adjoining owner with the apportionment depending on the use which the two owners make of the wall as well as the responsibility for the defect or disrepair.[580]

The conditional rights the building owner has, which are set out in section 2 of the PWA 1996, include underpinning, demolition, cutting into party structure and reducing height of a party wall or party fence wall.

The right to underpin under section 2(2)(a) of the PWA 1996 a party wall (fully or partially) relates to the need to strengthen the foundations of the party wall. Underpinning a party wall may be required in situations where there is settlement, subsidence or where additional weight is to be imposed on a party wall. This would be the case, for example, where new supports are fixed into the party wall, or an extension is added at the roof level. As a result, foundations may need to be extended so that the impact of the additional weight is spread further and deeper. Older buildings often have quite shallow foundations. Foundations of Victorian houses may rely on just three courses of corbelled brickwork. Underpinning is often undertaken in reinforced concrete. An adjoining owner must give its express consent before the building owner is allowed to place reinforced concrete over the boundary line onto the adjoining owner's property.[581]

As for subsidence, this refers to the ground underneath foundations giving way whereby it withdraws support from a party wall and causes it to drop. Where a party wall forms part of a building, such subsidence can put a strain on the building destabilising the walls. Some examples of causes behind subsidence include effects of droughts, tree roots (most severe effects are caused by poplars, oaks, willows, ash, plane and sycamore trees and fast-growing leylandii and eucalyptus) and heavy frosts which can cause clay sub-soils to shrink or swell severely. Other causes of subsidence include leaking drains, nearby excavation works (extension being erected next door), sinkholes and old mine-works.

Subsidence should not be mistaken for settlement, which is another type of downward movement. However, in most cases, buildings settle gradually over time in the process of the ground beneath being slowly compressed adjusting to the weight of the new building. This would apply, for example, where structural changes include a loft conversion. Period houses were built with very shallow foundations but were still able to accommodate a certain amount of movement. This is because foundations of period houses were built using flexible lime-based materials. Compared to modern concrete strip foundations of the 20th century, period houses were built with brick footings, which widened out at the base in a stepped pattern in order to be able to spread the load. The main reason why foundations have been built at a deeper level in the 20th century is to avoid movement in the ground caused by frost and seasonal moisture changes. Footings of period houses are less than half a meter deep and so their stability depends on the conditions of the ground they were built

---

580 A Redler, 'Practical Neighbour Law Handbook' (2007) RICS Books, 86.
581 ibid. 84.

on. While chalk and rock are firm, clay can dry out leading to shrinkages and cracks in hot weather, recovering later when swelling back into shape during wet weather. Where a party is concerned about subsidence it may decide to opt for underpinning, which effectively means making an excavation down to stable ground and pumping several tonnes of concrete into the excavations. This may however cause issues when applied to old buildings as it poses additional strain between the firm new area and the old walls. Another option, which is gentler and cheaper when it comes to period houses, is laying several courses of brick under the defective section of the party wall. In this way, contact is re-established with firm ground.

A method to strengthen a party wall includes thickening. This is more relevant to lower levels where the party wall carries an increased load caused by alterations to the attached building. Older buildings have usually thicker walls at lower levels than at upper levels. Where additional support is needed in relation to walls at upper levels, one can add brickwork to one side, which takes the line of the wall to the lower level. In this way, the load is carried and transferred down to the foundations throughout the thicker brickwork.[582]

The right to raise a party wall can extend to any purpose including raising a party wall at roof level to enclose a roof extension. It can also include raising a garden party fence wall to enclose on it for a new extension. The PWA 1996 gives no definition of the materials which should be used or the criteria how these should be selected for the purpose of raising a party wall. The optimal way however would be for the building owner raising the party wall to take reasonable requests of the adjoining owner into consideration, particularly in situations where the party wall will be visible only from the adjoining owner's property. Where the choice of materials is considered to be expensive, the cost can be shared by agreement between the building owner and the adjoining owner. In situations where building new brickwork directly on the old party wall below may not be considered prudent, where a party wall is old and in poor condition, there may be uncertainty about its ability to carry additional weight without the need to take the old party wall down and rebuild it. However, rebuilding a new party wall may also not be considered appropriate as it can result in substantial construction works.

A solution can be that the raised section of brickwork is supported on a cantilever from the frame of the new building (examples are set out in Diagrams 8.1 and 8.2 below). What this means is that the brickwork would overhang the full width of the party wall. While it would touch it at the junction, it would not impose weight on it. While this is considered to constitute raising a party wall under the PWA 1996 by most surveyors, there is some dissent as well. Therefore, each case should be evaluated on its own merit. This is an important consideration as, where the wall does not fall under the PWA 1996, there is no right of access to the land of the adjoining owner to build the wall.[583]

582 ibid. 84.
583 ibid. 84 and 85.

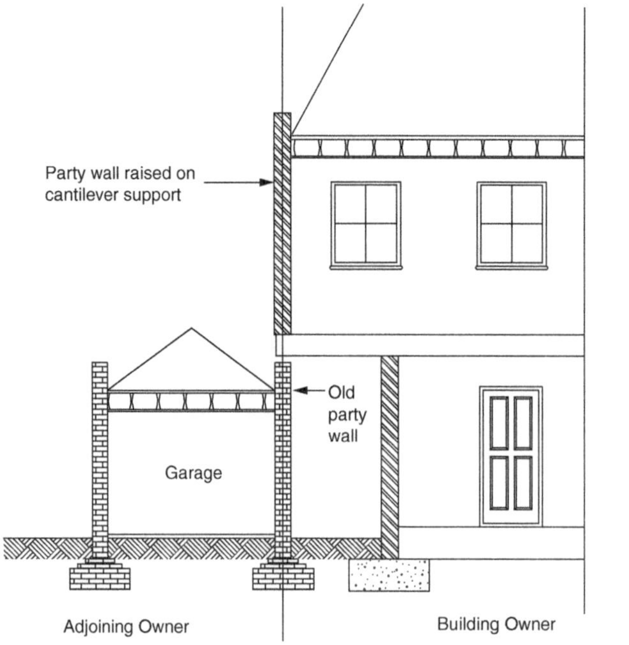

Diagram 8.1 New structure cantilevered fully over a party wall.[584]

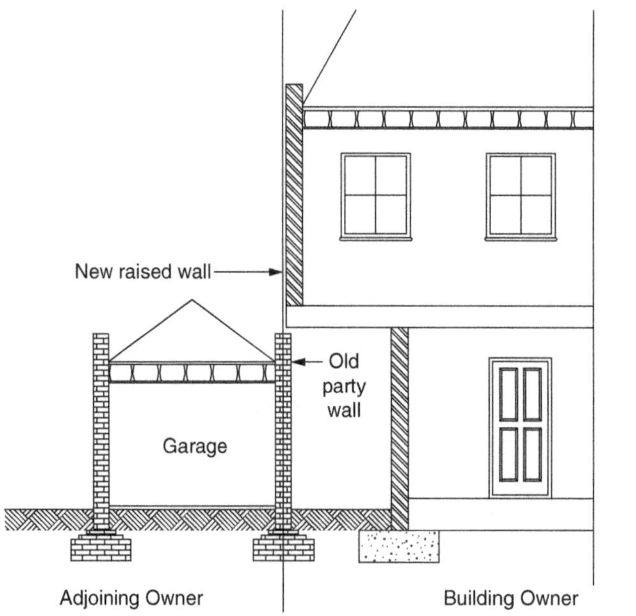

Diagram 8.2 New structure cantilevered over party wall, up to boundary line.[585]

584 ibid. 85.
585 ibid. 86.

The main purpose of raising or replacing a party fence wall is usually where a wall separates two gardens and is used to enclose a new extension. For example, a party fence wall needs to be raised to a height sufficient for the new extension. Equally, a party fence wall may need to be taken down and replaced with a new wall that is more substantial and meets modern regulations when enclosing a building.[586]

The right under section 2(2)(f) of the PWA 1996 of cutting into a party wall may become relevant where it is required to form a concrete pad stone to support a new steel beam, for example, where internal structural partitions are removed. Cutting into a party wall may also be required in order to strengthen roof joists in order to create a floor where a roof space is converted into a new room. The extent to which a cut is made into the party wall is not restricted by the PWA 1996. However, usually surveyors require that a concrete pad stone or beam does not extend further into the party wall than half the thickness of the wall. The reason for this is precautionary to ensure that the building owner's beam remains within his own land and does not encroach onto the adjoining owner's land or that the risk of damaging the adjoining owner's internal finishes is minimised. It also gives the adjoining owner the option to use an equal half thickness of the wall in that location. However, these restrictions are not rooted in the PWA 1996, and there may be situations where such restrictions should not apply. For example, this is the case where a house has very thin party walls and the building owner needs to insert the pad stone to a greater depth for the purposes of proper construction. Cutting into a party wall may also be needed where a chemical damp-proof course needs to be inserted or for the purposes of cutting in a flashing. There is also a right to cut away any projection from a party wall which includes footings, chimney breasts and flues. A party has the right to cut off any structure overhanging the party wall, which extends to pipes, gutters, eaves details as well as major structural cladding where a building has been built with parts of the elevation overhanging the party wall. As long as these structures overhang the party wall, they can be cut off. The level to which they overhang the party wall can be very small, and this still gives the right to cut them off.[587]

The right to reduce height of a party wall or party fence wall under section 2(2)(m) of the PWA 1996 gives the building owner the right to either reduce or demolish and rebuild at lower height party wall or party fence wall. The wall cannot be reduced to a height of less than two meters. In the case where the wall encloses the building of the adjoining owner, it must not be reduced to less than the height to which it is enclosed (with appropriate parapet). The adjoining owner has the option to serve a counter notice within 14 days of the original notice where the adjoining owner does not want to have the wall reduced in height. The notice would then require the building owner to maintain the current height of the party wall. However, in such a scenario, the adjoining owner has to then pay a proportion of the cost of the party wall to the building owner where it is either above two meters in height or above the adjoining owner's building's height. However, where the building owner incurs no cost (where, in order to comply with the counter notice, the wall is left in place and not reduced), the courts are likely to decide that the responsibility for maintenance as to the upper level of brickwork in the future will be with the adjoining owner. This would mean that the adjoining owner takes responsibility for the party wall as a result of the counter notice.[588]

586 ibid. 87.
587 ibid. 86–87.
588 ibid. 87.

A party wall can be exposed by construction works. This can happen, for example, where an initially internal face of the party wall is left exposed to the weather after an existing building has been demolished. Such an exposure can be permanent or temporary. In such a case, the PWA 1996 will apply and that will be the case even though no works have been performed on the party wall itself. Under section 2(2)(n) PWA 1996, there is an obligation for the building owner to ensure that an adequate weatherproofing to the wall is provided.[589]

In *Hughes v Percival*,[590] the claimant was responsible for considerable works to his property (Diagram 8.3). The works affected party walls on both sides and by the end of the construction works, independent contractors negligently damaged one of the party walls. This resulted in a substantial collapse damaging also the second party wall adjoining the defendant's property. It was held that the claimant was liable for the damage that the defendant suffered. This is because, when carrying out construction works to party walls, a building owner owes a non-delegable duty towards an adjoining owner. The claimant had a common law right to carry out the construction works to the party wall. However, this included a high risk of damage to the property of the adjoining owner (the defendant).

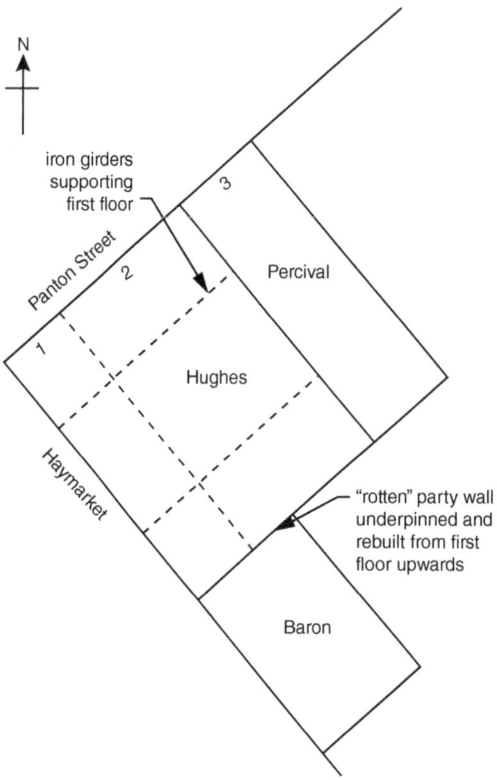

Diagram 8.3 *Hughes v Percival*.[591]

589 ibid.
590 (1883) 8 App Cas 443.
591 Chynoweth (n 276) 170.

Where a retaining wall collapses onto the neighbouring owner's land, it is the landowner from where the problem originated that has the obligation to clean up the debris and make sure that the issue will not occur again. This is the case even where no fault is shown. If an adjoining owner secures an *ex parte* injunction preventing a developer from breaking the terms of a party wall award, the adjoining owner will be entitled to costs if it is found that it was reasonable in seeking the injunction.[592]

In *Ward v Coope*,[593] a wall was built by a landowner's predecessors. The purpose of the party wall was to hold back the land from the neighbouring land. This is because the surface of the landowner's land was initially higher by about four feet than the neighbouring land. The difference in height increased to nine feet a few decades later. In 2010, there was a heavy snowfall because of which the wall collapsed onto the neighbouring land. There were no obvious warning signs prior to the collapse that would indicate the collapse was imminent. The Court of Appeal held that it was the landowner who bore the full responsibility as there was a mutual measured duty of care. Although a breach of the duty of care was not possible to establish, the landowner was responsible for the collapse arising from the addition of earth made by the predecessors. For the purposes of carrying out any engineering calculations and to do the work needed, the landowner had the right to enter the neighbour's land.

As a side note, while there was found to be an easement of support in relation to the four feet of higher land over the neighbouring land, as there was a change imposing a greater burden that cannot be separated from the original rights, the initial easement was destroyed. As there was no easement, the building owner was responsible as it was from its property where the issue arose. In this case it was held that the question was not 'if' the wall would collapse but 'when' the wall would collapse. An engineering survey helps in similar scenarios to allocate responsibility before a disaster ensues, but most importantly it helps prevent such disasters from occurring.[594] When considering structural issues in relation to party walls, section 2 of the PWA 1996 should be consulted as to the related rights and obligations of the building owner and adjoining owner so as to manage dispute avoidance measures. These will include having an open line of communication between the building and adjoining owners and managing any costs/losses linked to structural issues caught by section 2 of the PWA 1996.

---

592 'Boundaries: issues with neighbours' (2015) 35(10) *Property Law Bulletin*, 592, https://uk.westlaw.com/Document/I84B9E430D4E211E4A204A332330F1FAE/View/FullText.html?transitionType=SearchItem&contextData=(sc.Search).
593 [2015] EWCA Civ 30.
594 ibid.

9

Conclusion

By considering and actioning alternative dispute resolution avenues in situations where the PWA 1996 does not apply, where an award has been appealed in court or where ADR is better suited to parties' specific situation, members of the public have the potential to save money and time in relation to party wall disputes, and courts' workload would also be alleviated. This book aims to create a party wall legal area of its own merit, as a subset of neighbourly matters, with a holistic view of such disputes, by taking into account other areas of law and fact extending beyond the PWA 1996. By setting out links between such different legal areas as well as factual issues, it becomes possible to identify most common issues arising in the context of party walls so that they become predictable and therefore manageable. This in turn results in greater transparency, efficiency and effectiveness in party wall dispute prevention, management and resolution. This book provides different perspectives when it comes to party wall disputes, and in some cases even offers a choice of which areas of law/issues to prioritise to achieve a more favourable dispute outcome. It also provides suggestions for policy and legislative improvements.

The aim is to offer coherence, providing an overview of a range of factors that need to be taken into consideration when it comes to party wall disputes. It is to serve as an inspiration to investigate alternative methods of dispute resolution, to preserve neighbourly relationships and to save cost and time for members of the public.

There is an appetite for a consolidated and comprehensive legal understanding of party wall dispute resolution management. It is important that this area develops further, is fully recognised and continues to flourish in the years to come. This understanding should include effective and efficient dispute prevention, management and resolution methods, resulting in alleviating courts' heavy dispute workload and saving cost and time.

The multifaceted nature of party wall disputes calls for a reform of the party wall dispute resolution system as a whole. This book is aiming for a contextual approach to party walls as a valid area, where different areas of law and fact are brought together and through providing a comprehensive overview of issues linked to party walls to assist in a more cost- and time-effective party wall dispute prevention, management and resolution. The competing issues all play an important role in party wall disputes and need to be carefully considered when assessing what is the best strategy and dispute resolution route to take.

Party Wall Disputes: The Party Wall etc. Act 1996 and Beyond – *Legal Coherence and Dispute Management*, First Edition. Dr Laura Lintott.

As party walls have the potential to be highly personal in nature, it is common for disputes emanating out of them to grow acrimonious with emotions running high, standing in the way of practical, time- and cost-effective solutions. The analysis in this book shows different angles of party wall disputes by way of case law examples highlighting what types of problems can arise in the context of party walls. In order to be able to aid those with party wall issues, first there needs to be clarity in terms of what qualifies as a party wall, and secondly, there needs to be a defined legal area of party walls of its own merit. This means, taking into account different pieces of legislation regulating party-wall-related issues and disputes, case law interpreting such legislation clarifying how it applies in specific factual scenarios, the law of easements and the law torts.

It is this contextual analysis of the various aspects that can affect party walls that allows for a carefully carved out area of party walls and related disputes to be defined, which in turn leads to providing a selection of dispute resolution options and strategic advice to parties to party wall disputes, including precautions to be taken pre-dispute, dispute prevention, management and resolution options taking into account the parties' objectives, time and costs involved in such a dispute resolution process.

The new approach and findings in this book are reassuring in that they provide a rounded view of issues most likely to occur in connection with party wall disputes, showing a pattern of the types of problems linked to party walls. Where there is a pattern, there is a potential for problem-solving and systematisation of main recurrent issues assisting in navigating through party wall disputes.

In the common law system, there are very few legal specialists who are focussing on party wall disputes. The tendency is to qualify into a specific area, such as commercial law, family law, construction law, property law etc., focussing closely on the detail of each of these areas, which can lead to not seeing connections to other legal areas and the context of a multi-disciplinary problem. The concern is then the quality and consistency of surveyors' awards and legal advice given to parties to party wall disputes. By creating a party wall legal area of its own, members of the public that are at the pre-dispute stage or in the middle of a party wall dispute will benefit from consistency, cost and time efficiency, clarity and having more control over the strategy and consequences of their actions and dispute resolution choices. The lack of regulation of party wall surveyors puts an additional strain on party wall dispute resolution consistency.

It is the contextual analysis of party wall disputes that provides for a novel approach, whereby it is not just the PWA 1996 that stands in the centre of the issues governing them. On the contrary, this book explores areas of law that complement the PWA 1996 and are relevant where the PWA 1996's realm does not reach. While the PWA 1996 is a crucial piece of legislation in relation to party walls, it is but one piece of the puzzle concerning party wall disputes.

This book analyses and assimilates seven main issues.

First, it dissects the definition of what a 'party wall' is. An understanding of this is an important foundation that the rest of the book builds upon because the definition of 'party wall' is not isolated only to the wording of the PWA 1996. Instead, there are other pieces of legislation, as discussed in Chapter II, that aid the definition as well as DCLG's Explanatory Booklet.[595] In this context, Chapter II also considers works that can affect

595 (n 26).

party walls (for example, excavations) and what consequences these can have in relation to the relevant parties' rights and obligations, as well as dispute resolution mechanisms and procedures.

Secondly, this book analyses the PWA 1996, the history behind it, its purpose, the issues it covers and the rights and obligations of parties connected to party walls. The PWA 1996 is a pivotal piece of legislation in the context of party walls. However, its scope is limited, meaning that the rest of the book ventures well beyond the PWA 1996 and into different legal and factual areas providing contextualisation of the party wall area.

Thirdly, the book explores ways in which to manage and resolve party wall disputes. It elaborates on the mechanism in the PWA 1996, which provides for a notice dispute resolution mechanism but fails to provide other practical dispute resolution options. Chapter IV therefore goes beyond the PWA 1996 into litigation (while binding, this often provides a cumbersome and expensive solution that may not be practical for the parties due to the cost and time involved) and ADR methods, as well as dispute prevention, with a focus on the practical needs of the parties involved. This book promotes ADR mechanisms and dispute prevention, leading to more practical and digestible solutions for the parties, taking into account the fact that parties to party wall disputes often need to continue living in close proximity to each other for years or generations to come, meaning that preserving the relationship can be of value to them as well as limiting the cost and time of the party wall dispute.

Fourthly, this book brings together other pieces of legislation that affect party wall disputes and related rights and obligations, in order to provide a solid legislative platform where different statutes either complement the PWA 1996, overlap with it, or simply serve as a monitor to ensure that the PWA 1996 is compliant with the overall system of legislative rules within the society nationally (such as the Human Rights Act 1998).

Fifthly, having discussed the various legislative and dispute resolution aspects, this book delves into another area of law, namely proprietary rights in land affecting party walls and related disputes, with a particular focus on easements. The reason is because, as case law examples in Chapter VI show, it is not uncommon for party walls and easements to be intertwined. In such a situation, when a party wall dispute arises, its outcome and/or works performed on a party wall can have a knock-on effect on a linked easement. One has to be sensitive to easements in the context of party walls. The PWA 1996 is very clear on neither condoning nor authorising interferences with easements. While explaining which proprietary rights are particularly relevant to party walls, this book uses case law examples as to why that is and what the courts' inclination is when attempting to solve related party wall disputes and what factors play an important role. Party wall disputes, where easements are intertwined with a party wall or works performed on it, must be managed in such a way as to ensure that any such linked easements are not interfered with. If they are interfered with, the liable party should be aware of the consequences of its actions, such as, for example, a court ordering it to pay damages to the other party, or injunctions to be imposed upon it. Awareness of these issues means that parties can decide before a dispute arises what works to perform on a party wall and in what way, in order not to interfere with any connected easement with the aim of avoiding a party wall dispute altogether.

Sixthly, this book assesses torts with a focus on common law nuisance relevant to party walls, again using case law examples to demonstrate which tortious issues can be encountered in party wall disputes, and what the courts' tendency is when approaching such issues. Similarly to the area of proprietary rights in land, tortious issues (such as, noise, vibration or interference with light) can be intrinsically linked to party wall matters that impact such tortious rights. When a party wall dispute arises (and ideally before), one has to consider whether any infringements affecting tortious rights are likely to or have occurred. This will help parties to be aware of the consequences of their actions should related disputes arise and, for example, sanctions be ordered by a court.

Finally, it is not only different legal aspects and areas that are brought together in this book that relate to party wall disputes. Practical aspects and matters of fact, also play a non-negligible role in party wall issues, which is why this book discusses structural issues pertinent to party walls.

While the contextual overview of this book is one of the main new approaches, another factor that is of importance is the way in which disputes could or should be conducted in relation to party walls and the avenues available to the parties involved with the aim of a solution which is both practical and cost/time effective. While the surveyor procedure under the PWA 1996 is useful, it is not all-encompassing in terms of the variety of legal issues that can arise in the context of party walls. The PWA 1996 notice dispute resolution mechanism also does not always lead to a final satisfactory outcome. When appealing an award in court, while court proceedings are usually a possibility, the view formed in this book is that parties should at least consider the option of ADR when it comes to party wall disputes. There are a number of factors discussed that the parties need to take into account in order to find the most practical way forward in terms of dispute resolution strategy and management. Party wall disputes can grow acrimonious very quickly due to the personal nature of the properties involved. It is therefore important to perform a balancing act between the actual financial value in dispute, and any cost and time linked to the court procedure avenue. Where communication between the parties has broken down, a third party, for example a mediator, may be a potential and effective solution depending on the circumstances.

This book has analysed aspects that are relevant to party walls with the aim of providing an overview of the key issues from a practical perspective, where the parties are most likely to require legal advice or support as to party wall disputes. It does not aim to encompass every single problem that can stem from party walls, however it discusses an overview of the most pertinent areas that should be considered when discussing legal implications that can arise out of party walls and related disputes. To date most literature linked to party walls focusses mainly on the PWA 1996. While this book also covers aspects of the PWA 1996, one of its main aims is to venture well beyond the PWA 1996 – exploring other legal and factual areas that are highly relevant to party walls. It shows that a contextual approach leads to a more comprehensive grasp of party wall issues, providing a clearer overview to assist in identifying relevant issues and managing disputes linked to party walls. The only sensible and proper way to approach party wall problems is by viewing them in the context of the relevant legislation, proprietary rights, torts, practicalities of building practice and surrounding property, construction and commercial law. This book also promotes the importance of ADR combined with a practical and strategic approach, often leading to pragmatic solutions and a greater cost and time effectiveness.

Learning lessons from different legal and factual areas relevant to party walls, and from relevant reasoning and interpretation, forms the 'bricks' of an independent 'party wall area' on its own merit and can benefit party wall surveyors, legal practitioners, scholars as well as parties affected by party wall issues.

Parties to party wall disputes to date have very little to lean on. The PWA 1996 on its own is not comprehensive enough to give them all the protections and dispute resolution options that are out there. Beyond the PWA 1996, there is a chaos of different legal and factual issues that the parties do not necessarily anticipate or notice to start with. This can lead to negative outcomes, which could have been prevented by being aware of the main issues impacting on the relevant dispute, and by an appropriate cost- and time-sensitive dispute resolution strategy. This book has identified most pertinent legal and factual issues that can affect a party wall dispute and has set out what dispute resolution options parties can choose from to minimise their losses.

There is an appetite for a consolidated and comprehensive legal understanding of the party wall dispute management area from academic, professional (legal and surveyor) and client perspectives. It is important that this area develops further, is fully recognised and continues to flourish in the years to come including effective and efficient dispute prevention, management and resolution methods.

Bibliography

1. Legislation (Primary and Secondary) and Government Guidance

1.1 Access to Neighbouring Land Act 1992
1.2 Agricultural Holdings Act 1986
1.3 Agricultural Tenancies Act 1995
1.4 Arbitration Act 1996
1.5 Arbitration Act 2025
1.6 Bristol Improvement Act 1847
1.7 British Transport Commission Act 1945
1.8 British Waterways Act 1995
1.9 Building Act 1984
1.10 Building Regulations 1985 (SI 1985/1065)
1.11 Building Regulations 2010 SI 2010/2214
1.12 Building Research Establishment Guidelines
1.13 Channel Tunnel Rail Link Act 1996
1.14 Civil Liability (Contribution) Act 1978
1.15 Civil Procedure, The White Book Service 2024 (Sweet & Maxwell, 2024)
1.16 Contracts (Rights of Third Parties) Act 1999
1.17 Control of Pollution Act 1974
1.18 County Courts Act 1984
1.19 Crossrail Act 2008 (2008 c 18)
1.20 Department for Communities and Local Government, '*Party Wall etc. Act 1996: Explanatory Booklet*' (May 2016)
1.21 Electronic Communications Act 2000
1.22 Environmental Health Act 1990
1.23 Environmental Protection Act 1990
1.24 European Convention for the Protection of Human Rights and Fundamental Freedoms agreed by the Council of Europe at Rome on 4 November 1950
1.25 Fires Prevention (Metropolis) Act 1774 14 Geo 3 c LXXVIII
1.26 Hansard HL Debates Vol 568 31 Jan 1996, col 1536
1.27 High Speed Rail (London – West Midlands) Act 2017

Party Wall Disputes: The Party Wall etc. Act 1996 and Beyond – *Legal Coherence and Dispute Management*, First Edition. Dr Laura Lintott.

1.28 High Speed Rail (West Midlands - Crewe) Act 2021

1.29 HM Government, 'Resistance to the passage of sound – Approved Document E', 2003 edn Incorporating 2004, 2010, 2013 and 2015 Amendments (E1 – Protection against sound from other parts of the building and adjoining buildings; E2 – Protection against sound within a dwelling-house etc.; E3 – Reverberation in the common internal parts of buildings containing flats or rooms for residential purposes; and E4 – Acoustic conditions in schools), https://www.scribd.com/document/470048059/E-resistance-to-sound-2010-amend-2015

1.30 HM Land Registry Practice Guide 77, 'Altering the register by removing land from a title plan', https://www.gov.uk/government/publications/altering-the-register-by-removing-land-from-a-title-plan-pg77/practice-guide-77-altering-the-register-by-removing-land-from-a-title-plan

1.31 Hobart Building Act 1918 (9 Geo V No 75) (Tas)

1.32 House of Lords Debate, 31 January 1996, Vol 568 cc1535-48, https://api.parliament.uk/historic-hansard/lords/1996/jan/31/party-wall-bill-hl

1.33 House of Lord Debate, 22 May 1996, Vol 572 cc931-55, https://api.parliament.uk/historic-hansard/lords/1996/may/22/party-wall-bill-hl

1.34 Human Rights Act 1998

1.35 Land Registration Act 1925

1.36 Land Registration Act 2002

1.37 Land Registration Rules 2003/1417

1.38 Launceston Building Acts 1854 and 1894 (18 Vict No 18, 58 Vict No 32), (Tas)

1.39 Law Commission, 'Impact Assessment of the Law of Property Bill [2011]', June 2011

1.40 Law Commission, 'Making Land Work: Easements, Covenants and Profits a Prendre' (2011) 327, https://www.lawcom.gov.uk/project/easements-covenants-and-profits-a-prendre/

1.41 Law Commission, 'Reforming the law – Rights to Light' 356, 24, https://assets.publishing.service.gov.uk/government/uploads/system/uploads/attachment_data/file/391684/44872_HC_796_Law_Commission_356_PRINT.pdf

1.42 Law of Property Act 1925

1.43 Law of Property Bill

1.44 London Building Act 1894 (amended in 1905) 57 & 58 Vict. c CCXIII & 5 Edw. VII c CCIX respectively

1.45 London Building Act 1930 20 & 21 Geo 5 c CIVIII

1.46 London Building Acts (Amendment) Act 1939 2 & 3 Geo 6 c XCVII

1.47 London Government Act 1963

1.48 London Rebuilding Act 1667

1.49 Magistrates' Court Act 1980

1.50 Melbourne Building Act 1849 (13 Vict No 39) (Vic)

1.51 Metropolitan Building Act 1855 18 & 19 Vict c CXXII

1.52 Ministry of Housing, Communities & Local Government, 'Guidance – Preventing and Resolving Disputes in Relation to Party Walls' (2016), https://www.gov.uk/guidance/party-wall-etc-act-1996-guidance

1.53 Party Wall etc. Act 1996

1.54 Party Wall etc. Act 1996 (Commencement) Order 1997, SI 1997/670

1.55 Party Wall etc. Act 1996 (Electronic Communication) Order 2016 (SI 2016/335)

1.56 Planning (Listed Buildings and Conservation Areas) Act 1990
1.57 Pre-Action Protocol for Construction and Engineering Disputes (2nd edition)
1.58 Prescription Act 1832
1.59 Property Boundaries (Resolution of Disputes) Bill 2019–20
1.60 Rights of Light Act 1959
1.61 Senior Courts Act 1981
1.62 Sydney Building Act 1879 (42 Vict No 25) (NSW)
1.63 Technology and Construction Court Guide (TCC Guide)
1.64 Town and Country Planning Act 1990

2. Case Law

2.1 *Abbahall Ltd v Smee* [2002] EWCA Civ 1831
2.2 *Acco Properties Ltd v Severn* [2011] EWHC 1362 (Ch)
2.3 *Alan Wibberley Building Ltd v Insley* [1999] 1 W.L.R. 894
2.4 *Alcock (Douglas Alexander) v Wraith* [1991] 12 WLUK 156
2.5 *Aldin v Latimer Clark Muirhead & Co* [1894] 2 Ch. 437
2.6 *American Cyanamid Co v Ethicon Ltd* [1975] A.C. 396
2.7 *Anchor Brewhouse Developments Ltd v Berkley House (Docklands Developments) Ltd* (1988) 4 Const. L.J. 29
2.8 *Andreae v Selfridge & Co Ltd* [1938] Ch. 1
2.9 *Arena Property Services Ltd v Europa 2000 Ltd* [2003] EWCA Civ 1943
2.10 *Attorney-General v Sheffield Gas Consumers Co* (1853) 43 E.R. 119
2.11 *Barras v Aberdeen Steam Trawling & Fishing Co Ltd* [1933] AC 402
2.12 *Barry v Minturn* [1913] A.C. 584
2.13 *Baxter v Camden LBC (No. 2)* [2001] Q.B. 1
2.14 *Beaumont Business Centres Ltd v Florala Properties Ltd* [2020] EWHC 550 (Ch)
2.15 *R. (on the application of Best) v Chief Land Registrar* [2015] EWCA Civ 17
2.16 *Bickett v Morris* (1866) L.R. 1 Sc & Div. 47
2.17 *Bond v Nottingham Corporation* [1940] Ch. 429
2.18 *Brace v South East Regional Housing Association* [1984] 1 EGLR 144, CA
2.19 *Bradburn v Lindsay* [1983] 2 All E.R. 408
2.20 *Bridgland v Earlsmead Estates Ltd* 2015 3953034
2.21 *Broder v Saillard* (1876) 2 Ch. D. 692
2.22 *Browne v Flower* [1911] 1 Ch. 219
2.23 *Burlington Property Co Ltd v Odeon Theatres Ltd* [1939] 1 K.B. 633
2.24 *Butt v Imperial Gas Co* (1866-67) L.R. 2 Ch. App. 158
2.25 *Cambridge Water Co Ltd v Eastern Counties Leather Plc* [1994] 2 A.C. 264
2.26 *Cameron v Boggiano* [2012] EWCA Civ 157
2.27 *Carter v Cole* [2006] EWCA Civ 398
2.28 *Charalambous v Welding* [2009] EWCA Civ 1578
2.29 *Chartered Society of Physiotherapy v Simmonds Church Smiles* [1995] 1 E.G.L.R 155
2.30 *Chaturachinda v Fairholme* (CCCL, 23 September 2015)
2.31 *Chliaifchtein v Wainbridge Estates Belgravia Ltd* [2015] EWHC 47 (TCC)
2.32 *Clifford v Holt* [1898] 12 WLUK 27

2.33 *Colls v Home & Colonial Stores Ltd* [1904] A.C. 179
2.34 *Cook v JD Wetherspoon Plc* [2006] EWCA Civ 330
2.35 *Coventry (t/a RDC Promotions) v Lawrence* [2014] UKSC 13
2.36 *Crofts v Haldane* (1866-67) L.R. 2 Q.B. 194
2.37 *Crowley (t/a Crowley Civil Engineers) v Rushmoor BC* [2009] EWHC 2237
2.38 *Dalton v Henry Angus & Co* (1881) 6 App. Cas. 740
2.39 *Dawoodi v Zafrani* [2015] EWHC 3168 (TCC)
2.40 *Dean v Walker* (1997) 73 P. & C. R. 366, CA
2.41 *Dillard v F & C Commercial Property Holdings Ltd* [2014] EWHC 1219 QB
2.42 *Ann and Margaret Dodd v Samuel and James Holme* 110 E.R. 1296
2.43 *Drury v Army & Navy Auxiliary Co-operative Supply* [1896] 2 Q.B. 271
2.44 *Eastwood v Ashton* [1915] A.C. 900
2.45 *Edwards v Kumarasamy* [2016] UKSC 40
2.46 *Farrs Lane Developments Limited v Bristol Magistrates' Court* [2016] EWHC 982 (Admin)
2.47 *Fishenden v Higgs and Hill* (1935) 153 L.T. 128
2.48 *Frederick Betts Ltd v Pickfords Ltd* [1906] 2 Ch. 87
2.49 *Gaw v Coras Iompair Eireann* [1953] 1 WLUK 125
2.50 *Gillon v Baxter* [2003] EWCA Civ 1591
2.51 *Goldman v Hargrave* [1967] 1 A.C. 645
2.52 *Grant v Edmondson* [1931] 1 Ch. 1
2.53 *Gray v Elite Town Management Ltd* [2016] 11 WLUK 91
2.54 *Gyle-Thompson v Wall Street Properties Ltd* [1974] 1 W.L.R. 123
2.55 *Halsall v Brizell* [1957] Ch 169
2.56 *Hatton v United Kingdom (36022/97)* (2003) 37 E.H.R.R. 28
2.57 *Heathcote v Doal* [2017] 1 WLUK 613
2.58 *Heslop v Burns* [1974] 1 W.L.R. 1241
2.59 *Higgins v Betts* [1905] 2 Ch. 210
2.60 *Hiscox Syndicates Ltd v Pinnacle Ltd* [2008] EWHC 145
2.61 *Histed v Prosperity Developments Ltd* [2013] EWHC 4463 (QB)
2.62 *Hobbs Hart & Co v Grover* [1899] 1 Ch. 11
2.63 *Hobday v Nicholl* [1944] 1 All E.R. 302
2.64 *Hopgood v Brown* [1955] EWCA Civ 7
2.65 *Hough v Annear* [2007] 11 WLUK 31
2.66 *Hughes v Percival* (1883) 8 App Cas 443
2.67 *Hunt v Peake* (1860) John. 705
2.68 *Hunter v Canary Wharf Ltd* [1997] A.C. 655
2.69 *Ingram v Morecraft* (1863) 33 Beav. 49
2.70 *James v United Kingdom (A/98)* (1986) 8 E.H.R.R. 123
2.71 *James E McCabe Ltd v Scottish Courage Ltd* [2006] EWHC 538 (Comm)
2.72 *John Trenberth Ltd v National Westminster Bank Ltd* (1980) 39 P. & C.R. 104
2.73 *Jolliffe v Woodhouse* (1894) 10 TLR 553
2.74 *Jones v Pritchard* [1908] 1 Ch. 630
2.75 *Kaye v Lawrence* [2010] EWHC 2678 (TCC)
2.76 *Kelliher v Ash Estates Holding Ltd and another* (CLCC, 15 July 2013)

2.77 *Kempston v Butler* (1861) 12 I.C.L.R. 516
2.78 *Knight v Goulandris* [2018] 1 W.L.R. 3345
2.79 *Knight v Pursell* (1879) 11 Ch. D. 412
2.80 *Lea Valley Developments Ltd v Derbyshire* [2017] EWHC 1243 (TCC)
2.81 *Lea Valley Developments Ltd v Derbyshire* [2017] EWHC 1353 (TCC)
2.82 *Leakey v National Trust for Places of Historic Interest or Natural Beauty* [1980] Q.B. 485
2.83 *Llewellyn v Earl of Jersey* [1843] 11 M. & W. 183
2.84 *London and Manchester Assurance Co v O & H Construction* (1990) 6 Const. L.J. 155
2.85 *London, Gloucestershire & North Hants Dairy Co v Morley & Lanceley* [1911] 2 K.B. 257
2.86 *Longmire v Maldura* (CLCC, 29 January 2015)
2.87 *Longstaff International Ltd v Evans (Costs)* [2005] EWHC 4 (Ch)
2.88 *Lopez Ostra v Spain (A/303-C)* (1995) 20 E.H.R.R. 277
2.89 *Lotus Ltd v British* Soda Co Ltd [1972] Ch 123
2.90 *Lough v First Secretary of State* [2004] 1 W.L.R. 2557
2.91 *Louis v Sadiq* (1997) 74 P. & C. R. 325
2.92 *Lyle v Richards* (1866) L.R. 1 H.L. 222
2.93 *Carleton (Earl of Malmesbury) v Strutt & Parker (A Partnership)* [2008] EWHC 424 (QB)
2.94 *Marchant v Capital & Counties* [1983] 1 WLUK 326
2.95 *Matania v National Provincial Bank Ltd* [1936] 2 All E.R. 633
2.96 *Matos E Silva Lda v Portugal* (1997) 24 E.H.R.R. 573
2.97 *Maxted v Plymouth Corp* [1957] 3 WLUK 80
2.98 *Midland Bank Plc v Bardgrove Property Services Ltd* (1993) 65 P. & C.R. 153
2.99 *Miller v Jackson* [1977] 3 W.L.R. 20
2.100 *Mills v Blackwell* (1999) 78 P. & C.R. D43
2.101 *Mills v Savage* (CCCL, 15 June 2016)
2.102 *Mills & Rockleys Ltd v Leicester City Council* [1946] K.B. 315
2.103 *Mohamed & Lahrie v Antino & Stevens* (CCCL, 13 December 2017)
2.104 *Neilson v Poole* (1969) 20 P. & C.R. 909
2.105 *Nelson's Yard Management Co v Eziefula* [2013] EWCA Civ 235
2.106 *Network Rail Infrastructure Ltd v Williams* [2018] EWCA Civ 1514
2.107 *Observatory Hill Ltd v Camtel Investments SA* [1997] 1 E.G.L.R. 140
2.108 *Onigbanjo v Pearson* [2008] B.L.R. 507
2.109 *Ottercroft Ltd v Scandia Care Ltd* [2016] EWCA Civ 867
2.110 *Ough v King* [1967] 1 W.L.R. 1547
2.111 *Parkinson v Reid* [1966] 1 WLUK 544
2.112 *Patel v Peters* [2014] EWCA Civ 335
2.113 *Peck v Sims* CH-2022-000132
2.114 *Phipps v Pears* [1965] 1 QB 76
2.115 *Pomfret v Ricroft* 85 E.R. 454
2.116 *Power v Shah* [2023] EWCA Civ 239
2.117 *Presland v Bingham* [1889] 3 WLUK 16
2.118 *Proudfoot v Hart* (1890) 25 Q.B.D. 42

2.119 *Prudential Insurance Lt v Waterloo Real Estate Inc* [1999] EWCA Civ J0122-13
2.120 *Pyer v Carter* 156 E.R. 1472
2.121 *Ray v Fairway Motors (Barnstable)* (1969) 20 P. & C.R. 261
2.122 *Raymond v Young* [2015] EWCA Civ 456
2.123 *Redland Bricks Ltd v Morris* [1970] A.C. 652
2.124 *Reed Executive Plc v Reed Business Information Ltd* [2004] EWCA Civ 887
2.125 *Rees v Skerrett* [2001] 1 W.L.R. 1541
2.126 *Reeves v Blake* [2010] 1 W.L.R. 1
2.127 *Reeves v Young Young & Antino* (CCCL, 3 January 2017)
2.128 *Regency Villas Title Ltd v Diamond Resorts (Europe) Ltd* [2018] 3 W.L.R. 1603
2.129 *Reinhardt v Mentasti* (1889) 42 Ch. D. 685
2.130 *Roadrunner Properties Ltd v Dean* [2003] EWCA Civ 1816
2.131 *Sack v Jones* [1925] Ch. 235
2.132 *Scott v Firth* 176 E.R. 595
2.133 *Scratton v Brown* (1825) 4 B. & C. 485
2.134 *Seeff v Ho* [2011] EWCA Civ 186
2.135 *Selby v Whitbread & Co* [1917] 1 K.B. 736
2.136 *Sheffield Masonic Hall Co Ltd v Sheffield Corp* [1932] 2 Ch. 17
2.137 *Shelfer v City of London Electric Lighting Co (No. 1)* [1895] 1 Ch. 287
2.138 *Smith v Smith* (1875) L.R. 20 Eq. 500
2.139 *Southwark LBC v Mills* [2001] 1 A.C. 1
2.140 *Southwark LBC v Tanner* [2001] 1 A.C. 1
2.141 *Standard Bank of British South America v Stokes* (1878) 9 Ch. D. 68
2.142 *Dewdney Stedman and Smith Stedman v Henry Attwell Smith* (1857) 8 El. & Bl. 1
2.143 *Stepney Corp v Gingell Son & Foskett Ltd* [1909] A.C. 245
2.144 *Arbitration between Stone and Hastie, Re* [1903] 2 K.B. 463
2.145 *Stran Greek Refineries v Greece (A/301-B)* (1995) 19 E.H.R.R. 293
2.146 *Takhar v Mohamed* [2023] EWHC 2190 (KB)
2.147 *Targett v Ferguson* (1996) 72 P. & C.R. 106
2.148 *Taylor v Jones* [2024] EWCA Civ 170
2.149 *Taylor v Whitehead* (1781) 2 Doug. K.B. 745
2.150 *Udal v Dutton* [2007] EWHC 2862 (TCC)
2.151 *R. (on the application of Vella) v Lambeth LBC* [2006] Env. L.R. 33
2.152 *Video London Sound Studios Ltd v Asticus (GMS) Ltd and Keltbray Demolition Ltd* [2001] 3 WLUK 133
2.153 *Walby v Walby* [2013] 1 P. & C.R. DG6
2.154 *Ward v Coope* [2015] EWCA Civ 30
2.155 *Watson and others v Croft Promo-Sport Ltd* [2009] EWCA Civ 15
2.156 *Wellington Properties Ltd v The Trustees of the Will of the Second Duke of Westminster* [2018] EWHC 3048 (Ch)
2.157 *Welter v McKeeve* (CCCL, 27 November 2018)
2.158 *West Faulkner Associates v Newham LBC* (1995) 11 Const. L.J. 157
2.159 *Weston v Arnold* (1872-73) L.R. 8 Ch. App. 1084
2.160 *Wigginton & Milner Ltd v Winster Engineering Ltd* [1978] 1 W.L.R. 1462
2.161 *Wilson v First County Trust Ltd* [2004] 1 A.C. 816

2.162 *Woodhouse v Consolidated Property Corp Ltd* [1992] 11 WLUK 28
2.163 *Yorkshire Water Services Ltd v Sun Alliance and London Insurance Plc* [1998] Env. L.R. 204
2.164 *Zissis v Lukomski* [2006] EWCA Civ 341

3. Books

3.1 A Walsh, *A Practical Guide to Neighbour Disputes and the Law* (Law Brief Publishing, 2020)
3.2 C Sara, D Dovar, *Boundaries and Easements*, 7th edn, (Sweet & Maxwell, 2019)
3.3 D Cavill, S Wheeler, Professor M Dixon, *Ruoff & Roper, Registered conveyancing* (Sweet & Maxwell, 2021)
3.4 J Gaunt QC, The Hon Mr Justice Morgan, *Gale on the Law of Easements*, 21st edn, (Sweet & Maxwell Ltd, 2020)
3.5 N Isaac and M Hearsum, *The New Party Wall Casebook* (Property Publishing, 2019)
3.6 N Isaac, *The Law and Practice of Party Walls*, 2nd edn (Property Publishing, 2020)
3.7 P Chynoweth, *The Party Wall Casebook* (Blackwell Publishing Ltd, 2007)
3.8 Pyramus & Thisbe Club, *The Party Wall Act Explained – A Commentary on the Party Wall Etc. Act 1996*, 3rd edn (2016)
3.9 R Megarry, W Wade, *The Law of Real Property*, 9th edn, (Sweet & Maxwell Ltd, 2019)
3.10 S Bickford-Smith, A Francis, T Weekes, *Rights of Light, The Modern Law* 3rd edn (Jordan Publishing, 2015)
3.11 S Bickford-Smith, D Nicholls, A Smith, *Party Walls Law and Practice* 4th edn (Lexis Nexis, 2017)
3.12 The Hon. Sir V Ramsey, S Furst QC, *Keating on Construction Contracts*, 11th edn, Mainwork & 3rd Supplement, (Sweet & Maxwell Ltd, 2023)

4. Articles, Notes, Guidance

4.1 A Moor, 'Property litigation blog; Boundary disputes', *Practical Law UK Articles*, Resource ID W-008-6296 (11 July 2017), https://uk.practicallaw.thomsonreuters.com/w-008-6296?transitionType=Default&contextData=(sc.Default)&firstPage=true
4.2 A Redler, *Practical Neighbour Law Handbook* (RICS Books, 2007)
4.3 B Spenser, https://pwbsl.com/party-wall-surveyor/party-wall-surveyor/
4.4 'Boundaries: issues with neighbours' (2015) 35(10) *Property Law Bulletin*, 592. https://uk.westlaw.com/Document/I84B9E430D4E211E4A204A332330F1FAE/View/FullText.html?transitionType=SearchItem&contextData=(sc.Search)
4.5 Building Research Establishment, Site layout planning for daylight and sunlight: A guide to good practice' (2022), https://images.reading.gov.uk/2022/11/ID95-Site-Layout-Planning-for-Daylight-and-Sunlight-A-Guide-to-Good-Practice-BRE-2022.pdf
4.6 C Cant, 'Easements – Adjoining properties and party walls' (2013). https://www.christophercant.co.uk/wp-content/uploads/2020/09/Easements-Adjoining-properties-and-party-walls.pdf

4.7 C Parker-Jones, 'Sound testing to comply with Part E for residential developments', https://www.parkerjonesacoustics.com/insights/articles/sound-testing-explained?_sm_au_=isVqLCN6nnV6w05VQcLJjKQ1j7GJ1Colliers

4.8 CIArb, 'What is dispute resolution?', https://www.ciarb.org/dispute-services/what-is-dispute-resolution/

4.9 'Common law nuisance' (Maintained), *Practical Law Environment*, Resource ID 6-502-4804, https://uk.practicallaw.thomsonreuters.com/Document/Ib5554812e83211e398db8b09b4f043e0/View/FullText.html?transitionType=SearchItem&contextData=(sc.Search)#co_anchor_a698923

4.10 'Easements: point of access' (1999) 20(3) *Property Law Bulletin* 18–19

4.11 F Langdon, I Buller, 'Party wall insulation and noise from neighbours' (1977) 55(4) *Journal of Sound and Vibration* 495–507

4.12 H Carr, 'Grenfell Tower and the failure of building and fire safety regulations introduction' (2017) 20(5) *Journal of Housing Law* 110–115

4.13 Harvey – Norman Architects, 'Party Wall Act of 1996 – What is it and what are your obligations under it?', https://www.harveynormanarchitects.co.uk/articles/the-party-wall-act-of-1996-definitions-and-obligations

4.14 HHJ E Bailey, 'Proper conduct for the party wall surveyor', A Faculty of Party Wall Surveyors Webinar – 12 November 2020

4.15 Hogan Lovells, 'Practice & law – developing themes' (2013) *The Estate Gazette, ABI/INFORM Collection*, 120

4.16 HS2, 'Party wall guidance under the HS2 regime' (2022), www.hs2.org.uk

4.17 M Edwards, 'Fire safety in tall buildings — Part 1: tall, super-tall and mega-tall' (2017) 33(7) *Construction Law Journal* 442–445

4.18 M Richards, N Roberts, R Abbey, S Desmond, 'When is the covenantor liable to carry out repairs under his repairing covenant?' (1996) *Butterworths Property Law Service Case Comment*

4.19 M Rousell, 'Is a wall enclosing a railway embankment a party wall?' *Practical Law UK, Ask*, Resource ID 5-522-4375 (8 July 2015), https://uk.practicallaw.thomsonreuters.com/5-522-4375?originationContext=document&transitionType=DocumentItem&contextData=%28sc.Default%29&comp=pluk

4.20 N Isaac, 'What can party wall surveyors learn from arbitration? (Part 1)' (2019) *Party Wall & Boundaries Hub*, http://www.boundariesbook.co.uk/?p=985

4.21 P Aliker, 'Party walls and arbitration' (2005) 155 *New Law Journal* 866

4.22 P Chynoweth, 'Invalid party wall awards and how to avoid them' (University of Salford Manchester, 2000), https://salford-repository.worktribe.com/output/1457272/invalid-party-wall-awards-and-how-to-avoid-them

4.23 P Chynoweth, 'The Party Wall etc Act 1996: Compensation and treatment of easements', (2004) *The Conveyancer and Property Lawyer*, 232–240

4.24 Party Wall Company, 'Party wall process for adjoining owners', http://partywallcompany.co.uk/PDF/The_Party_Wall_Process_For_Adjoining_Owners.pdf

4.25 Party Wall Company, 'Party wall process for building owners', http://partywallcompany.co.uk/PDF/The_Party_Wall_Process_For_Building_Owners.pdf

4.26 'Party Wall Mediation Scheme LLP', https://www.partywallmediation.co.uk

4.27 'Party Wall etc Act 1996 – appeal procedure under section 10(17)' (2006) 22(7) *Construction Law Journal*, 469–481.

4.28 Practical Law Construction, 'PWA 1996: reviewing the party wall surveyor's award' (Maintained) *Practical Law UK Practice Note*, Resource ID 7-516-5148, https://uk.practicallaw.thomsonreuters.com/7-516-5148?originationContext=document&transitionType=DocumentItem&contextData=%28sc.Default%29&comp=pluk

4.29 'Procedure: obligations to the court' (2016), 37(6) *Property Law Bulletin*, 44, https://uk.westlaw.com/Document/IED200F109B7C11E6BC4BD8CA38210D45/View/FullText.html?

4.30 R Pratt, '4 ways you might be affected by the Party Wall Act' (LexisNexis, 2013)

4.31 RICS, *Party Wall Legislation and Procedure – England and Wales*, 7th edn, (RICS, 2019)

4.32 RICS professional guidance, UK, *Rights of Light* (2nd edn, 2016)

4.33 Royal Borough of Kensington and Chelsea, 'Code of construction practice: Minimising the impact of noise, vibration and dust' (2016), https://www.rbkc.gov.uk/environment/advice-builders/code-construction-practice-minimising-impact-noise-vibration-and-dust

4.34 S Bickford-Smith, C Lamont, *Party Wall etc. Act 1996 – 10 years on* (Landmark Chambers, 2007)

4.35 S Bickford-Smith, H Davies, 'Bankers' bunkers: neighbours' rights' (2014) *Estates Gazette*

4.36 S Radcliffe, 'Measuring noise nuisance' Resource 0-502-1969 (2010)

4.37 S Tozer, G Fetherstonhaugh QC, J Karas QC, N Cheffings, M Ditchburn, 'Protocol for disputes between neighbours about the location of their boundary (The Boundary Disputes Protocol)' (2017), https://www.propertyprotocols.co.uk/the-boundary-disputes-protocol

4.38 S Waldron, 'Crossrail: lifting legislative provisions – Easing the load' (RICS isurv, 2017), https://www.isurv.com/info/390/features/11162/crossrail_lifting_legislative_provisions

4.39 T Reid, 'The Party Wall etc. Act 1996 (PWA 1996)' (Maintained) *Practical Law UK Practice Note*, Resource ID 8-383-5739, https://uk.practicallaw.thomsonreuters.com/Document/Ib5556c44e83211e398db8b09b4f043e0/View/FullText.html?transitionType=SearchItem&contextData=(sc.Search)#co_anchor_a952869

Index

Note: *Italic* page numbers refer to *figure.*

Party Wall Disputes: The Party Wall etc. Act 1996 and Beyond – *Legal Coherence and Dispute Management*, First Edition. Dr Laura Lintott.